Adaptive Array Measurements in Communications

For a complete listing of the *Artech House Antennas and Propagation Library*, turn to the back of this book.

Adaptive Array Measurements in Communications

M. A. Halim

Artech House
Boston • London
www.artechhouse.com

Library of Congress Cataloging-in-Publication Data
Halim, M. A.
    Adaptive array measurements in communications / M. A. Halim.
       p.   cm. — (Artech House antennas and propagation library)
    Includes bibliographical references and index.
    ISBN 1-58053-278-0 (alk. paper)
    1. Adaptive antennas.   2. Antenna arrays.   3. Radio—Interference.
    4. Wireless communication systems.   I. Title.   II. Series.
  TK7871.67.A33   H35   2001
  621.382'4—dc21

                                                 00-068933
                                               CIP

British Library Cataloguing in Publication Data
Halim, M. A.
    Adaptive array measurements in communications. — (Artech House antennas and
    propagation library)
    1. Antenna arrays   2. Radio—Interference   3. Wireless communications systems
    I. Title
    621.3'824

    ISBN 1-58053-278-0

Cover design by Gary Ragaglia

© **2001 ARTECH HOUSE, INC.**
685 Canton Street
Norwood, MA 02062

International Standard Book Number: 1-58053-278-0
Library of Congress Catalog Card Number: 00-068933

10 9 8 7 6 5 4 3 2 1

To Nadia and Skander
Love
Dad

Contents

Preface

Although there are many books available on the subject of adaptive arrays, they usually contain rigorous mathematical analysis involving matrix and vector algebra. However, it is not necessary to use all of that complex mathematics just to acquire a general knowledge about the principle of operation of an adaptive array system. Moreover, a reasonably good understanding of the subject itself can also be obtained using only simple conventional mathematics and a few basic matrix relations found in many books of standard mathematical tables.

In this book, an attempt has been made to predict the degree of cancellation of the undesired signals by an adaptive array system. An attempt has also been made to describe the adaptive array principles, at the systems level, in a very simple manner using very simple mathematics.

In Chapter 1, the concept of the principle of adaptive array is developed by analyzing its basic operations with the aid of a simple analytical model.

In Chapter 2, the theoretical capabilities of an adaptive array system to remove the undesired signal from the received signal, under various operating conditions, are discussed.

In Chapter 3, analysis of an adaptive array system is presented using simple matrix relations where matrix expressions have been related to linear expressions to reveal their physical significance.

Finally, appendixes are provided that give final output power derivations for different types of systems as well as tables of phase values for cross-coupled transmission lines.

This book is especially suitable for students at the postsecondary levels, in college or university, who are interested in acquiring some basic knowledge in the field of adaptive arrays but do not wish to get involved with complex mathematics.

The idea of this book came to me a while ago during the preparation of a technical report on the subject at the Canadian Marconi Company, Kanata, Ontario, Canada. It was felt that a book that presented a general description of the adaptive array system and included a step-by-step guide to the principle of cancellation of undesired signals and the mathematical roots of the subject would be useful.

I am greatly indebted to Mr. Michael Roper, director of engineering, SR Telecom, Kanata, Ontario, Canada, for his many constructive and valuable suggestions during the preparation of this book.

Introduction

One of the most damaging factors associated with any radio communication network is the effect of interference signals. These signals fall within the desired operating frequency band and therefore can cause considerable degradation in system performance, including complete annihilation of the desired signal. These interference signals could be the desired signal itself arriving with different phase shifts due to multipath propagation or they could be generated by some other communication systems located within the range and operating inside the same frequency band. They could also be generated by some unfriendly party attempting to jam the desired communication link. Therefore, from both commercial and military points of view, it is desirable to find some means to eliminate or at least reduce the effect of these undesirable interference signals.

In general, to effectively remove the interference signals, the receiver would ideally have a primary input terminal which will bring in the desired signal contaminated by interference signals and a secondary input terminal which will bring in only the contaminating or interference signals. Then, at the front end of the receiver, the contaminating signals could simply be subtracted from the contaminated signal to recover the desired signal.

A common radio receiver could operate as this type of ideal receiver if the associated antenna is replaced by a beam former whose input terminals are connected to a linear array. This array beam former assembly has the ability to direct signals, arriving from different directions, to different output ports of the beam former.

A beam former has an equal number of input and output ports so that an $N \times N$ beam former will have N input ports and N output ports. The

N input ports are connected to the elements of an N-element linear array. If a signal generator is now connected to an output port, then a beam will be formed in one and only one specific direction. Thus, N number of signal generators connected to N output ports will produce N number of beams in N different specific directions. A beam former can be designed so that it will produce a beam along the array boresight called the *main beam,* while the rest of the beams, called the *auxiliary beams,* will be symmetrically located on both sides of the main beam. If along the peak of any given beam all the other beams have nulls, then the beams are said to be mutually orthogonal.

Again, a distant signal source located along the peak of a given beam will produce a signal only at the output port corresponding to that particular beam. Thus, a signal located along the main beam will appear only at the main output port while that along one of the auxiliary beams will appear only at the corresponding auxiliary output port.

Note that besides the linear configuration, other array structures could also produce orthogonal beams. This could be verified by measuring the beam patterns produced by any array structure. Then, if a given beam produces nulls along the peaks of all the other beams, then the beams are mutually orthogonal.

To obtain the desired signal, free of interference, the array is oriented such that the main beam is pointed toward the desired signal source while the interference signal sources are scattered on both sides of the boresight. Then, only the main output port will carry the desired signal. On the other hand, if none of the interference signal sources lie along any of the auxiliary beams, then all the output ports, main and auxiliary, will carry those signals. Thus, the signal from each auxiliary port could be properly weighted and subtracted from the main port signal to obtain the desired signal, free of interference.

The use of a beam former could, however, be omitted if one of the array elements is a high-gain antenna, termed the *main antenna,* and the others are relatively low-gain antennas, termed the *auxiliary antennas.* Then the output from each auxiliary antenna could be properly weighted and subtracted from the main antenna output to reduce the interference signals carried by it. A system of this type is also known as a sidelobe canceller.

In practice, the interference signals are automatically removed from the main port signal. To achieve this, all the output ports of the beam former are connected to a signal processing network where each auxiliary port is connected to a feedback loop, which will contain the main port signal and the weighted signal from the auxiliary port. When subjected to an appropriate algorithm, the weighter would adjust itself to generate the correct amplitude and phase to cancel part of the interference signals present in the main output port. Thus, if all the loops associated with all the auxiliary ports are activated simultaneously, then, theoretically, all the interference signals contained in the main port signal could be cancelled. This arrangement is known as *adaptive signal processing* and the array–beam former assembly, along with the signal processing network, is known as an adaptive array. Thus, the signal appearing at the output port of an adaptive array is the desired signal, free of interference, ready to be applied at the input port of a radio receiver.

The adaptive signal processing could utilize either an analog or digital technique. In the past, an LMS algorithm has been used almost exclusively in conjunction with the feedback loops. However, in recent days, although LMS algorithm is still widely used, the RLS technique is becoming increasingly popular in digital applications. Many excellent books and papers have been written on the subject of adaptive signal processing techniques, which is beyond the scope of this book.

This book assumes an active adaptive array system is already available. Then the theoretical capabilities of the system to remove the interference signals are calculated under various operating conditions. It also discusses briefly the complex mathematical expressions, involving matrix and vector algebra, found in related textbooks, and presents them in much simpler forms to establish their physical significance. In addition, some key devices and subsystems used in the construction of an adaptive array are discussed.

1

Fundamental Concepts

Introduction

An adaptive array system can be divided into a number of basic subsystems. Each subsystem is designed to perform a specific function. A complete system is obtained by cascading these basic subsystems. Thus, when a desired signal, contaminated by undesired signals and noise, is applied at the input end of the system, a noise-free pure desired signal emerges from its output end.

In this chapter, various primary subsystems of an adaptive array are identified and their functions discussed. The presentation has been made in a relatively simple manner for easy understanding by a beginner.

The operation of a typical adaptive array system has been explained with the help of a simple basic model. Simple mathematical equations are derived. Then the development and construction of a basic adaptive system, using these equations, are demonstrated.

1.1 Conceptual Representation of Adaptive Arrays

Figure 1.1 shows a conceptual representation of an adaptive array. It contains an N-element linear dipole array connected to the N input ports of an $N \times N$ orthogonal beam former. The output of the beam former is fed to a signal processing network, which includes, among others, splitters, combiners, feedback circuits, weighters, and multipliers. The signal processing network has one output port, which is connected to the input port of a radio receiver.

1

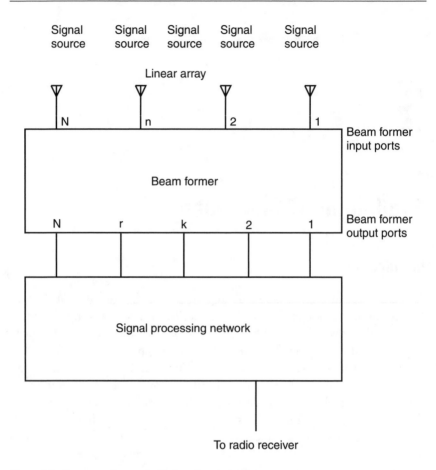

Figure 1.1 Conceptual representation of an adaptive array.

Then there are distant signal sources, transmitting both desired and undesired signals. Each of the array elements will intercept signals from all transmitting signal sources and deliver them to the corresponding input port of the beam former. The beam former will transport these signals to all of its output ports from which they will be delivered to the signal processing network, which will process them and deliver them to the radio receiver.

1.2 The Linear Array

A dipole array could have different geometrical configurations, such as a linear array, circular array, rectangular array, and so on. Here, only the linear array is considered.

1.2.1 Structure and Behavior of a Linear Array

An *N*-element linear array is shown in Figure 1.2. It contains *N* number of dipoles arranged in a row where the spacing between two adjacent dipole elements is *d*. For most practical applications, *d* is made equal to half the wavelength of the operating frequency. The direction perpendicular to the plane of the array, on either side of the plane, is called the *boresight*.

The array is driven by a feed network which has *N* number of ports at one side, designated antenna ports, and a single port at the other side,

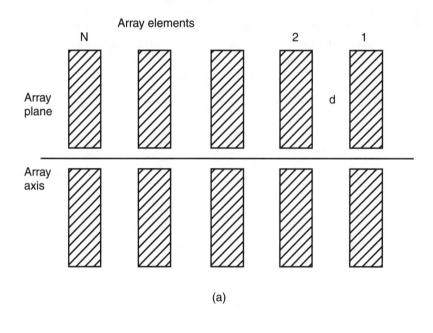

(a)

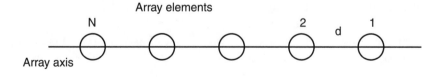

(b)

Figure 1.2 *N*-element linear array: (a) side view and (b) top view.

designated the receive port. The N number of array elements are connected to the N number of antenna ports of the feed network. Then, in the receive mode, a receiver is connected to the receive port while in the transmit mode, a signal generator is connected to this port. The antenna ports are usually called the *input ports* of the feed network and the generator or receiver port is called the *output port.*

Although a dipole is an omnidirectional antenna, a linear array of dipoles is directional where the direction of maximum radiation depends on the spacing of the array elements and also on the phase difference between the signals at two adjacent array elements.

For example, suppose the spacing d between two adjacent array elements is half wave at the operating frequency and a signal generator is connected to the output port of the feed network. Now, the feed network is adjusted such that the phase difference between the signals at two adjacent array elements is zero. Then the signals from all the array elements will be in phase along the array boresight and 180° out of phase along the ends of the array axis. Hence, the maximum radiation will be along the boresight and nulls along the both ends. An array with a beam along the boresight is called a *broadside array* [Figure 1.3(a)].

On the other hand, if the feed network is adjusted to make the phase difference between the signals at two adjacent array elements 180°, then the signals from all the array elements will be 180° out of phase along the boresight and in phase along the ends. Hence, the maximum radiation will be along the ends [Figure 1.3(b)] and nulls along the boresight. An array with beams along the ends is known as an *endfire array.*

Thus, with half-wave array spacing, the phase difference between the signals at two adjacent array elements could be made to lie anywhere between 0° and 180° by adjusting the feed network and, hence, the maximum radiation can be placed along any desired direction.

When the same array is used to produce a number of beams along different directions, the beam along the boresight is called the *main beam* while the others are called the *auxiliary* or *AUX beams.* It is customary to take the boresight as the reference so that the locations of all the other beams are measured by the angle $\theta°$ off the boresight. Thus, for the main beam $\theta = 0$ and for the rth AUX beam $\theta = \theta_r$ off the boresight.

As can be seen from Figure 1.3, the radiation pattern of a linear array is bidirectional and has the form of a figure eight. To make it unidirectional, a ground plane reflector is placed in one side of the array so that the beam in that side is reflected by the ground plane to combine with the beam in the other side to form a unidirectional radiation pattern. The space between

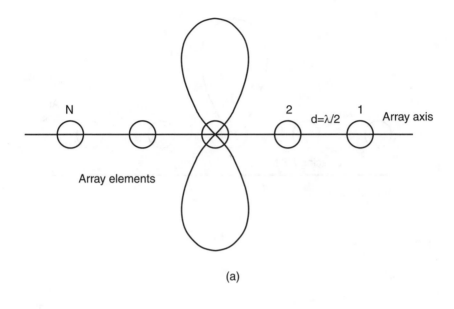

(a)

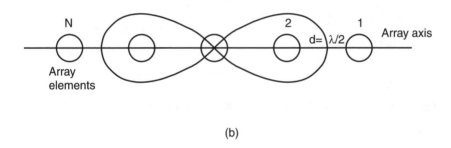

(b)

Figure 1.3 Linear array radiation pattern: (a) broadside array and (b) endfire array.

the ground plane could be anywhere between one-eighth and a quarter of a wavelength at the center frequency of the operating band [1]. This will not only remove the figure-eight ambiguity but also double the gain, except for the endfire case, where both the beams of the figure eight will remain and only the lower half of each beam (below the horizon) will be reflected back (Figure 1.4).

1.2.2 Array Radiation Formulas and Patterns

The wavefront, due to a distant signal source located along the array boresight, is parallel to the plane of the array or the array axis. Then, if the signal

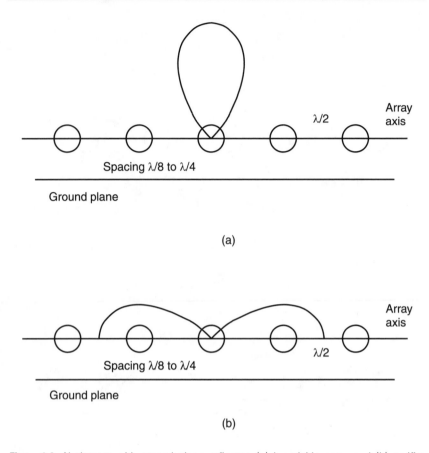

Figure 1.4 *N*-element with ground plane reflector: (a) broadside array and (b) endfire array.

source moves clockwise along a circle, the wavefront will rotate clockwise [Figure 1.5(a)]. Similarly, if the signal source moves counterclockwise along a circle, the wavefront will rotate counterclockwise [Figure 1.5(b)]. Although a full circle rotation of $+360°$ or $-360°$ is required to obtain the complete radiation pattern, due to symmetry, a half circle rotation of $+90°$ to $-90°$ or from $-90°$ to $+90°$ is usually sufficient.

Let the source move clockwise by $+\theta°$ off the boresight. Then the wavefront will rotate clockwise by $+\theta°$ [Figure 1.5(c)]. Hence, if d is the distance between two adjacent array elements, then the distance l_n of the nth array element from the wavefront is

$$l_n = (n - 1)d \sin \theta \qquad n = 1, 2, \ldots, N \qquad (1.1)$$

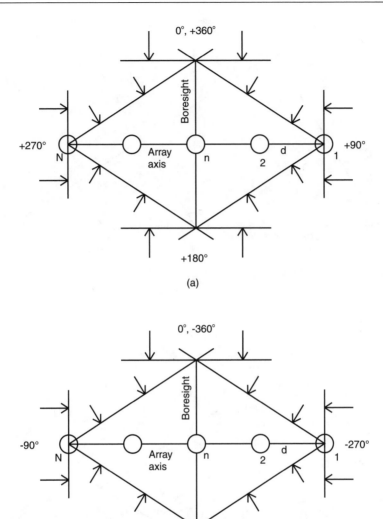

Figure 1.5 Incoming signals and wavefronts. The direction of incoming signals is indicated by an arrow and the wavefront by thick lines. (a) Clockwise rotation of the signal source or the wavefront. (b) Counterclockwise rotation of the signal source or the wavefront. (c) Path-length difference between element 1 and other array elements due to a source at $+\theta°$ off the boresight. (d) Path-length difference between element 1 and element n due to a source along $+\theta°$ off the boresight and along the rth beam at $+\theta_r°$ off the boresight.

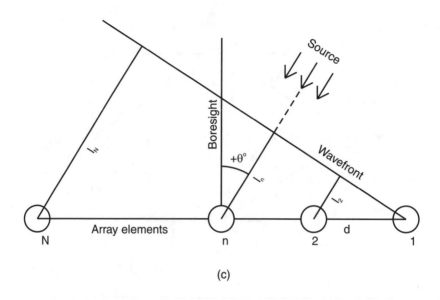

(c)

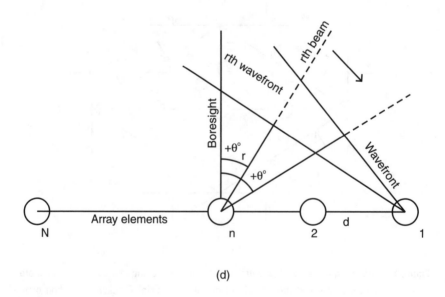

(d)

Figure 1.5 (continued).

and represents the excess path to be traveled by the signal from the source to reach element n, relative to element 1. Taking element 1 as the reference, the phase difference between element n and element 1 is given by

$$\psi_{n_1} = \psi_n = \beta l_n = (n-1)\beta d \sin\theta \qquad n = 1, 2, \ldots, N \qquad (1.2)$$

where $\beta = 2\pi/\lambda$ and λ is the wavelength of the operating frequency.

For element 1, $n = 1$ and hence, $\psi_{11} = \psi_1 = 0$ is the phase at element 1 and represents the reference phase. Also, from (1.2), the phase difference between any two adjacent array elements is

$$\psi_{n(n-1)} = \psi = \beta d \sin\theta \qquad (1.3)$$

As mentioned above, the spacing d between two adjacent array elements is equal to half the wavelength of the operating frequency. Thus, $d = \lambda/2$ and $\beta d = \pi$ and hence from (1.2) and (1.3), the phase ψ_n at the nth array element and the phase difference ψ between two adjacent array elements are given by

$$\psi_n = (n-1)\pi\sin\theta = (n-1)\psi \qquad n = 1, 2, \ldots, N \qquad (1.4a)$$

and $\qquad \psi = \pi\sin\theta \qquad (1.4b)$

For a counterclockwise rotation of the wavefront [Figure 1.5(d)], θ is negative and hence ψ is negative.

1.2.2.1 Signal Source Along the Main Beam

Let the feed network be adjusted so that the resulting beam is along the array boresight creating the main beam. Then the transfer function or gain between a signal source located along the boresight or the main beam and the output port of the feed network is given by

$$G_M(\theta) = \sum_{n=1}^{N} e^{jn\psi} = \sum_{n=0}^{N-1} e^{jn\psi} = \frac{\sin N(\psi/2)}{\sin(\psi/2)} e^{j(N-1)\psi/2} \qquad (1.5a)$$

Considering the magnitude only, the normalized transfer function or gain is given by

$$|G_M(\theta)| = \left|\frac{\sin N(\psi/2)}{N\sin(\psi/2)}\right| \qquad (1.5b)$$

1.2.2.2   Signal Source Along the rth AUX Beam

In this case the feed network is adjusted to produce a beam along $+\theta_r$ off the array boresight which could be designated as the rth AUX beam. Then the transfer function or gain $G_r(\theta)$ between a signal source located along the rth AUX beam and the output port of the feed network can be obtained by using a method similar to that for the main beam.

First, the position of the array is adjusted for the maximum received signal so that the signal source will be located along the peak of the rth AUX beam. Then, from (1.4b), the corresponding phase difference between two adjacent array elements is given by

$$\psi_r = \pi \sin \theta_r \tag{1.6a}$$

Now, let the source move clockwise or counterclockwise in a circle. But any angular location θ with respect to the boresight corresponds to a phase difference ψ between two adjacent array elements. Therefore, with respect to the rth beam, the phase difference ψ_R between two adjacent array elements is given by

$$\psi_R = \psi - \psi_r \tag{1.6b}$$

Then the normalized transfer function or gain between the source and the rth AUX port will be given by

$$|G_r(\theta)| = \left| \frac{\sin N(\psi_R/2)}{N \sin(\psi_R/2)} \right| = \left| \frac{\sin N[(\psi - \psi_r)/2]}{N \sin[(\psi - \psi_r)/2]} \right| \tag{1.7a}$$

Note that along the boresight, $\theta = \psi = 0$ and, hence, the gain of the rth AUX beam in the direction of the boresight is given by

$$|G_r(0)| = \left| \frac{\sin(N\psi_r/2)}{N \sin(\psi_r/2)} \right| \tag{1.7b}$$

1.3   The Beam Former

One of the main subsystems of an adaptive array is the beam former, which is discussed in detail in Section 2.13. A brief discussion is presented here.

1.3.1 The Definition and Function of a Beam Former

Basically, a beam former is a feed network that can feed an antenna array simultaneously from different signal generators to produce beams in different directions. It has an equal number of input and output ports. The ports connected to an antenna array are usually called the input ports while those connected to the receivers or signal generators are called the output ports.

A beam former is constructed in such a way that a signal generator, connected to a specific output port, will create a specific phase difference between two adjacent input ports. Hence, if a linear array is now connected to the beam former, then a beam will be formed along a specific direction corresponding to this phase difference.

Let a signal generator be connected to the rth output port of an $N \times N$ beam former; then the resulting phase difference ψ_r between two adjacent array elements will be given by

$$\psi_r = (2\pi/N)r \qquad r = 0, 1, 2, \ldots, (N-1) \qquad (1.8)$$

From (1.6a), this will produce a beam along θ_r off the array boresight, given by

$$\theta_r = \sin^{-1}(\psi_r/\pi) \qquad (1.9)$$

Thus, an $N \times N$ beam former, having N number of output ports, is capable of producing N number of beams along N number of specific directions.

A beam former is called *orthogonal* when, along the peak of a given beam, all the other beams have nulls.

1.3.2 Characteristics of Orthogonal Beam Formers

Consider the $N \times N$ orthogonal beam former shown in Figure 1.1 whose input ports are connected to an N-element linear array and the output ports are connected to a signal processing network.

Now, let the signal processing network be disconnected from the beam former and a signal generator be connected to a specific output port while the other output ports are match terminated. Then a beam will be formed in one and only one specific direction, relative to the array boresight. Conversely, if a signal source is placed along a specific beam, then the signal will appear at one and only one specific output port. Therefore, each output port of a beam former corresponds to a beam and each beam corresponds to an output port of the beam former.

One of the beams lying along the array boresight is called the main beam whose corresponding output port is called the main output port or the main port. The other beams are called the auxiliary or AUX beams and are located on both sides of the array boresight. Each of these AUX beams corresponds to an AUX output port or an AUX port. Therefore, if a signal source connected to the rth AUX port produces a beam along, say, $+\theta_r$ off the array boresight, then the signal from a source located along $+\theta_r$ will appear only at the rth AUX port.

For effective performance, a beam former is usually placed behind the ground plane used as the reflector for the array.

1.3.3 Separation of Desired and Undesired Signals

In an adaptive array system, the main beam is pointed toward the desired signal source while the interference signal sources are scattered on both sides of the main beam. They may or may not coincide with any of the AUX beams. Hence, the desired signal will appear only at the main port while an interference signal, whose source is located along a given beam, will appear only at the corresponding AUX port. All other interference signals, whose sources are not located along any beam, will appear at every AUX port as well as at the main port.

Thus, the main port will contain both the desired and the interference signals while all the AUX ports will contain only the interference signals and no desired signal. Hence, the AUX port signals could be combined with the main port signal, with correct amplitude and phase, to remove the interference signals from the main output port, leaving only the desired signal.

Accordingly, inside the signal processing network, each AUX port signal is passed through a weighter to provide it with the correct amplitude and phase and then combined with the main port signal by a combiner. The output of the combiner is the desired signal, free of interference, which appears at the output port of the signal processing network, ready to be delivered to a radio receiver.

1.3.3.1 Determination of the Locations of the Beams

Substituting (1.8) into (1.9), the location θ_r of the rth beam, relative to the array boresight, can be expressed as

$$\theta_r = \sin^{-1}(\psi_r / \pi) = \sin^{-1}(2r/N) \qquad r = 0, 1, 2, \ldots \qquad (1.10)$$

Thus, the location of any beam, relative to the boresight, can be easily calculated.

Example 1.1

Determine the locations of all the beams formed by a 4 × 4 beam former.

Answer

For a 4 × 4 beam former, $N = 4$ and, hence, (1.8) and (1.10) will give

$$
\begin{array}{lll}
\text{for } r = 0 & \psi_0 = 0 & \theta_0 = 0 \\
r = 1 & \psi_1 = 90° & \theta_1 = 30° \\
r = 2 & \psi_2 = 180° & \theta_2 = 90° \\
r = 3 & \psi_3 = 270° = -90° & \theta_3 = -30°
\end{array}
$$

Thus, a 4 × 4 beam former will produce four beams where one is along the boresight and the other three are along +30°, −30°, and ±90° (endfire) off the boresight, respectively.

Example 1.2

Prove that the signal from a source, located along an AUX beam, will not appear at the main output port.

Answer

The gain $G_r(0)$ of the rth AUX beam along the direction of the boresight is given by (1.7b). Then,

$$
\begin{array}{lll}
\text{for AUX 1 beam} & \psi_1 = 90° & G_1(0) = 0 \\
\text{AUX 2 beam} & \psi_2 = 180° & G_2(0) = 0 \\
\text{AUX 3 beam} & \psi_3 = 270° & G_3(0) = 0
\end{array}
$$

Thus, all the AUX beams have nulls along the array boresight and, hence, no signal from any of the AUX beams will appear at the main output port.

Example 1.3

Plot the patterns of the main beam and all the AUX beams for a 4 × 4 beam former.

Answer

First consider the main beam. Substituting (1.4b) into (1.5b), the normalized transfer function or gain, for $N = 4$, can be expressed in terms of θ given by

$$|G_M(\theta)| = \left| \frac{\sin 4[(\pi \sin \theta)/2]}{4 \sin[(\pi \sin \theta)/2]} \right| = \left| \frac{\sin(360 \sin \theta)}{4 \sin(90 \sin \theta)} \right| \qquad (1.11)$$

$G_M(\theta)$ can now be plotted as a function of θ by varying θ from $-90°$ to $+90°$.

Similarly, substituting (1.4b) and (1.6a) into (1.7a) gives the normalized transfer functions or gains for different AUX beams in terms of θ and hence can be plotted against θ by varying θ from $-90°$ to $+90°$. Figure 1.6 shows the patterns of all the four beams due to a 4×4 beam former. For clarity, the plots have been made with both positive and negative polarities. We can see from the figure that in the direction of a given beam, all the other beams have nulls.

1.3.4 Measurement of Array Radiation Patterns

1.3.4.1 Different Measurement Techniques

The beam patterns shown in Figure 1.6 could also be obtained experimentally by measuring the output signals at the main port and at each AUX port due to a distant radiating signal source. Alternatively, a signal source could be placed at the main port and at each AUX port while the radiated signal is measured by a distant receiver. During each measurement, the array is rotated 360° clockwise or counterclockwise.

Measurements could also be made by keeping the array fixed and moving the distant radiating signal source or the receiver in a circle, clockwise or counterclockwise, with the array at the center. In the fixed array setup, the wavefront at the array location will rotate, clockwise or counterclockwise, in the same way as the rotating array itself, as shown in Figures 1.5(a) and (b).

Thus, there are four possible alternate methods to obtain the beam patterns experimentally: a rotating array as a transmitter or a receiver or a fixed array as a transmitter or a receiver. For the purpose of demonstration, the method of a fixed array as the receiver is considered here.

1.3.4.2 Radiation Pattern Measurement with Fixed Array and Moving Source

As mentioned above, the method of the radiation pattern measurement considered here will be the one in which the array is kept fixed while the

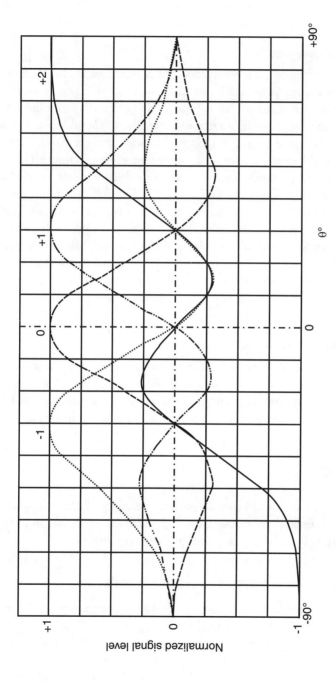

Figure 1.6 Beam patterns of a 4 × 4 beam former. (For clarity of the plots, both negative and positive polarities have been used.)

radiating signal source is moved in a circular path with the array as the center.

Main Beam Measurement

For the pattern $G_M(\theta)$, the received signal is measured at the main output port. First, the array is oriented such that the received signal is maximum, indicating that the signal source is located along the array boresight or the main beam and the received wavefront is parallel to the plane of the array.

Now, the signal source is moved clockwise along a circle. Then the wavefront would rotate clockwise as shown in Figure 1.5(a). Similarly, if the signal source is moved counterclockwise, the wavefront would rotate counterclockwise as shown in Figure 1.5(b). The full radiation pattern is obtained by moving the signal source through a full circle.

However, as mentioned earlier, due to symmetry, a half circle movement of ±90° by the source would provide information for a full circle pattern. A half circle movement of the signal source corresponds to ±90° rotation of the wavefront and is equivalent to a clockwise rotation of 0° to +90° and +270° to 360° [Figure 1.5(a)] or a counterclockwise rotation of 0° to −90° and −270° to −360° [Figure 1.5(b)].

Auxiliary Beam Measurement

For the pattern $G_r(\theta)$ of the rth AUX beam, the received signal is measured at the rth AUX port. The orientation of the array is adjusted such that the received signal is maximum, indicating that the signal source is located along the rth AUX beam. The pattern is now measured in the same way as $G_M(\theta)$.

1.4 The Signals

The radiating signal sources contain both the desired signal and the undesired signals. They are randomly located throughout the space so that an incoming signal could arrive at the array location from any direction.

1.4.1 Distribution of Signal Sources and Signals

As mentioned earlier, there are distant signal sources transmitting the desired signal $f_d(t)$ and interference signals $f_{i_1}(t), f_{i_2}(t), \ldots$ (Figure 1.7). These signal sources are distributed randomly across the space so that some of them may coincide with one or more beams of the beam former.

Each element of the array will pick up signals from all the signal sources, desired or undesired, and deliver them to its corresponding beam

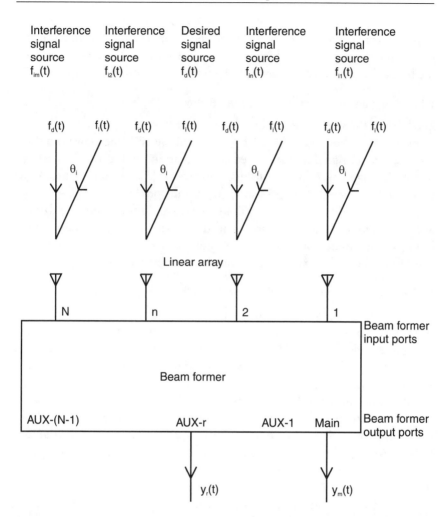

Figure 1.7 A two-source, two-port system showing multiple signal distribution.

former input port. Therefore, every input port of the beam former will carry signals from all the signal sources, desired or undesired.

The beam former will now transport the signals from all of its input ports to all of its output ports. However, the signal from a source located along a given beam will appear only at the corresponding output port. On the other hand, signals from sources not located along any of the beams will appear at every output port.

1.4.2 System with Two Transmitting Signal Sources

This is the simplest system configuration and will be studied here to help develop the basic concept. One of the signal sources is the desired signal source, located along the boresight or the main beam. The other is an interference signal source which could be located anywhere, either side of the boresight, within ±90°, representing one of the sources from Figure 1.7.

1.4.2.1 Signals at the Input Ports of the Beam Former

Let $f_d(t)$ be the signal being transmitted by the desired signal source located along the array boresight. Also, let $f_i(t)$ be the signal being transmitted by the interference signal source located along an arbitrary direction $+\theta_i$ off the boresight (Figure 1.7), where θ_i does not coincide with any of the AUX beams. It is assumed that $f_d(t)$ and $f_i(t)$ are uncorrelated.

Each of the array elements will intercept signals from both the desired and the interference signal sources and deliver them to the corresponding beam former input port. Hence, all input ports of the beam former will carry both the desired signal $f_d(t)$ and the interference signal $f_i(t)$.

1.4.2.2 Transfer Functions or Gains at the Active Output Ports

There are only two active output ports: the main port and an arbitrary rth AUX port. The rest of the output ports are match terminated. Consequently, only three transfer functions or gains will have to be considered: (1) between the desired signal source and the main output port, (2) between the interference signal source and the main output port, and (3) between the interference signal source and the rth AUX output port.

Let the transfer function or gain between the desired signal source and the main port be denoted by $G_M(0)$. Then $G_M(0)$ will represent the transfer function or gain between the desired signal source and all the input ports and between all the input ports and the main port.

Again, let $G_M(\theta_i)$ be the transfer function or gain between the interference signal source and the main port. Then $G_M(\theta_i)$ will represent the gain or transfer functions between the interference signal source and all the input ports and between all the input ports and the main port.

Similarly, let the transfer function or gain between the interference signal source and the rth AUX port be denoted by $G_r(\theta_i)$. Then, $G_r(\theta_i)$ will represent the gain or transfer functions between the interference signal source and all the input ports and between all the input ports and the rth AUX port.

1.4.2.3 Signals at the Output Ports of a Beam Former

Since the source of desired signal $f_d(t)$ is located along the main beam, the desired signal will appear only at the main port. On the other hand, the direction of the source of the interference signal $f_i(t)$ does not coincide with any of the AUX beams and so it will appear at the main port as well as at all the AUX ports. For a two-output system, all output ports, except the main port and the rth AUX port, are match terminated.

Now, the main output port contains the desired signal $f_d(t)$ and part of the interference signal $f_i(t)$. On the other hand, the rth AUX port contains only the interference signal $f_i(t)$ and no desired signal. Then at the main port, the signal $y_M(t)$ is given by

$$y_M(t) = G_M(0)f_d(t) + G_M(\theta_i)f_i(t) \qquad (1.12)$$

and at the rth AUX port, the signal $y_r(t)$ is given by

$$y_r(t) = G_r(\theta_i)f_i(t) \qquad (1.13)$$

Signals $y_M(t)$ and $y_r(t)$ will now be delivered to the signal processing network.

1.5 Signal Processing Network

The function of the signal processing network is to receive signals from all the active output ports of the beam former and process them to remove the undesired signals so that its output port contains only the desired signal to be delivered to a radio receiver.

1.5.1 Removal of Interference Signals by the Signal Processing Network

From (1.12) it is seen that the main port signal $y_M(t)$ contains the desired signal $f_d(t)$ along with the interference signal $f_i(t)$. Therefore, to obtain the desired signal, free of interference, the interference signal $f_i(t)$ must be removed from $y_M(t)$.

Now, (1.13) shows that the rth AUX port signal $y_r(t)$ contains only the interference signal $f_i(t)$. Therefore, the interference signal from $y_M(t)$ could be removed if $G_r(\theta_i)f_i(t)$ in (1.13) is made equal to $G_M(\theta_i)f_i(t)$ in (1.12) with opposite polarity. Then, when $y_r(t)$ is combined with $y_M(t)$ by a combiner, the two interference signal components will cancel each other, leaving only the desired signal $G_M(0)f_d(t)$ in the main port signal $y_M(t)$.

Thus, the combined signal at the output port of the combiner will contain only the desired signal. This canceling process is carried out inside the signal processing network and, hence, only the pure desired signal will appear at the output port of the signal processing network.

However, in order to make $G_r(\theta_i)f_i(t)$ equal to $G_M(\theta_i)f_i(t)$ with opposite polarity, the signal processing network must contain a weighter which has the capability of modifying the amplitude and phase of a signal passing through it. Then when the rth AUX port signal $y_r(t)$ is passed through this weighter, it will acquire the correct amplitude and phase to cancel the interference signal present in $y_M(t)$. The parameters of the weighter could be adjusted manually or it could also be adjusted automatically by a feedback circuit subjected to an appropriate algorithm.

Note that the device which measures and adjusts the weights of a signal passing through it is functionally a complex multiplier. However, we cannot call it a multiplier because that name has been assigned to another device in the circuit. Also, to avoid confusion with the quantity the device is supposed to measure and adjust, that is, the weight, we thought it better not to call it a weight either. Thus, in this book we use the term *weighter* to denote the device.

1.5.2 Final Output Signal

Figure 1.8 shows a signal processing network which contains a weighter W. The signal $y_r(t)$ from the rth AUX output port is applied at the input port of the weighter so that at its output port, the signal $y_r'(t)$ will be given by

$$y_r'(t) = Wy_r(t) = WG_r(\theta_i)f_i(t) \tag{1.14}$$

Signal $y_r'(t)$ is then combined with the signal $y_M(t)$ from the main port by the combiner Σ, to produce the total output signal $y_T(t)$ given by (Figure 1.8)

$$y_T(t) = y_r'(t) + y_M(t) \tag{1.15}$$
$$= WG_r(\theta_i)f_i(t) + G_M(0)f_d(t) + G_M(\theta_i)f_i(t)$$

This total output signal $y_T(t)$ will appear at the output port of the signal processing network as the final output signal $y_o(t)$, which will then be delivered to the input port of a radio receiver.

1.5.3 Final Output Power

From (1.15), the total output power P_T will be given by

$$P_T = |y_T(t)|^2 = |y_r'(t)|^2 + |y_M(t)|^2 + 2\,\mathrm{Re}[y_r'(t)y_M(t)] \tag{1.16}$$

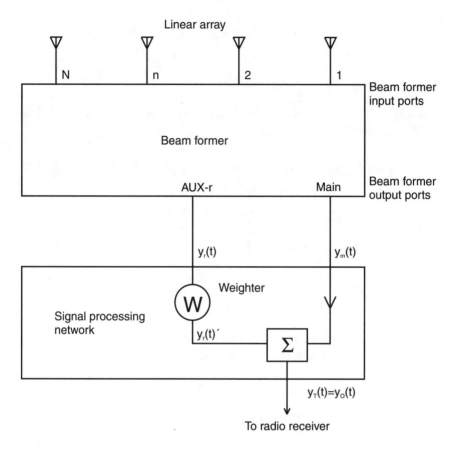

Figure 1.8 Two-input signal processing network with manually adjustable weighter.

Keeping in mind that $f_d(t)$ and $f_i(t)$ are uncorrelated,

$$
\begin{aligned}
|y_M(t)|^2 &= [G_M(0)f_d(t) + G_M(\theta_i)f_i(t)]^2 \\
&= |G_M(0)|^2|f_d(t)|^2 + |G_M(\theta_i)|^2|f_i(t)|^2 \\
&\quad + 2\,\mathrm{Re}[G_M(0)G_M(\theta_i)]\,[f_d(t)\,f_i(t)] \\
&= |G_M(0)|^2|f_d(t)|^2 + |G_M(\theta_i)|^2|f_i(t)|^2
\end{aligned}
\tag{1.17}
$$

But

$$
|y_r'(t)|^2 = W^2|G_r(\theta_i)|^2|f_i(t)|^2
\tag{1.18}
$$

and

$$[y_M(t)\,y_r'(t)] = [G_M(0)f_d(t) + G_M(\theta_i)f_i(t)]\,WG_r(\theta_i)f_i(t)$$
$$= W\,|\,G_M(0)G_r(\theta_i)\,|\,[f_d(t)f_i(t)] \tag{1.19}$$
$$+ W\,|\,G_M(\theta_i)G_r(\theta_i)\,|\,|\,f_i(t)\,|^2$$

Now $|f_d(t)|^2 = P_d$ = power of the desired signal and $|f_i(t)|^2 = P_i$ = power of the interference signal. Therefore, (1.17), (1.18), and (1.19) become

$$|y_M(t)|^2 = |\,G_M(0)\,|^2 P_d + |\,G_M(\theta_i)\,|^2 P_i \tag{1.20a}$$
$$|y_r'(t)|^2 = W^2\,|\,G_r(\theta_i)\,|^2 P_i \tag{1.20b}$$
$$[y_M(t)\,y_r'(t)] = W\,|\,G_M(\theta_i)G_r(\theta_i)\,|\,P_i \tag{1.20c}$$

Substituting (1.20) into (1.16) gives the total output power P_T as

$$P_T = |\,G_M(0)\,|^2 P_d + |\,G_M(\theta_i)\,|^2 P_i + W^2\,|\,G_r(\theta_i)\,|^2 P_i \tag{1.21}$$
$$+ 2W\,|\,G_M(\theta_i)G_r(\theta_i)\,|\,P_i$$

Equation (1.21) represents the final output power which could be measured at the output port of the signal processing network.

1.5.4 Cancellation of the Interference Signal

The mechanisms by means of which the undesired signals are removed are discussed next.

1.5.4.1 Principle of Cancellation

From (1.21), we can see that P_T is a function of W and hence will change with any change in W. However, the weighter is connected to the rth AUX port, which contains only the interference signals. Therefore, any change in weighter parameters could affect only the interference signals present in the total output power P_T while the desired signal power P_D present in P_T will not be affected.

Thus, any adjustment of the weighter parameters, causing P_T to decrease, would decrease only the interference signal present in P_T. Therefore, if we continue to adjust the weighter, then the interference signal present in P_T will keep on decreasing and hence P_T will keep on decreasing.

However, ultimately a stage will be reached at which all interference signals will be removed and P_T will contain only the desired signal P_D. At that stage, no further adjustment of the weighter will decrease P_T and a steady state will have been reached.

1.5.4.2 Manual Adjustment of the Weighter Parameters

In this case, manually adjustable weighters are used whose parameters are physically adjusted to obtain the minimum total output power. The process involved is shown in Figure 1.8 where the radio receiver will be replaced by a monitor so that the signal from the output port of the signal processing network, representing the total output power P_T, could be displayed on the monitor. The parameters of the weighter are then manually adjusted, and the display on the monitor is observed, until the display indicates the minimum. This will correspond to the minimum total output power and hence the removal of the interference signals.

From (1.21), we note that the variation of P_T with respect to W can be represented mathematically by

$$dP_T/dW = 2W|G_r(\theta_i)|^2 P_i + 2|G_M(\theta_i)G_r(\theta_i)|P_i \qquad (1.22)$$

Then, when P_T becomes minimum, a steady state is reached and the derivative of P_T becomes zero. If W_{SS} represents the weighter at which P_T becomes minimum and the steady state is reached, then (1.22) will give

$$dP_{T\min}/dW_{SS} = 2W_{SS}|G_r(\theta_i)|^2 P_i + 2|G_M(\theta_i)G_r(\theta_i)|P_i = 0 \qquad (1.23)$$
$$\text{or} \quad W_{SS} = -G_M(\theta_i)/G_r(\theta_i) \qquad (1.24)$$

Equation (1.22) gives the values of the weighter parameters at which the total output power P_T will become minimum and the interference signals will be removed.

Note, however, that although theoretically possible, the manual adjustment of the weighter parameters is not very practical. However, it is quite useful to understand the mechanism of operation. In a practical system, automatic adjustment of the weighter parameters is normally employed, in which case the system is also known as an adaptive system.

1.5.4.3 Automatic Adjustment of the Weighter

In this section, the principle and operation of automatically adjustable weighters are discussed.

Time Derivative of the Weighter

In this method, the weighter parameters, represented by W, are automatically adjusted to obtain the minimum value of the total output power P_T. An array system, equipped with the mechanism for automatic adjustment of the weighter, is also called an *adaptive array*.

The process therefore involves a self-adjusting weighter with a control input port and a feedback circuit controlled by an appropriate algorithm. Then, the feedback circuit will deliver part of the total output signal y_T (Figure 1.8) at the control input port of the weighter so that the weighter parameters will be adjusted for the minimum value of y_T and hence the minimum value of P_T. Thus, once the minimum value of P_T has been reached, the process of any further adjustment of W will cease. This will indicate that the weighter has reached its steady-state value, W_{SS}, and will remain constant from then on at that value. In other words, the time derivative of W_{SS} will become zero.

Now, as (1.23) indicates, at steady state, the weighter derivative of P_T becomes zero. Likewise, at steady state, the time derivative of the weighter also becomes zero. Hence, in (1.23), dP_{Tmin}/dW_{SS} can be replaced by dW_{SS}/dt to give

$$dW_{SS}/dt = 2W_{SS}|G_r(\theta_i)|^2P_i + 2|G_M(\theta_i)G_r(\theta_i)|P_i = 0 \quad (1.25)$$

Therefore, the general expression for the time derivative of the weighter could be written as

$$dW/dt = 2W|G_r(\theta_i)|^2P_i + 2|G_M(\theta_i)G_r(\theta_i)|P_i \quad (1.26)$$

which, at steady state, will become (1.25).

Alternate Representation of the Time Derivative

To implement the automatic adjustment of the weighter parameters, we need to develop the appropriate circuitry. However, to develop that circuitry, it is necessary to express (1.25) in a more convenient form that can be directly translated into the desired feedback circuit.

Let the AUX port output signal $y_r(t)$ be multiplied by the complex conjugate of the total output signal $y_T(t)$ to give

$$[y_r(t)y_T^*(t)] = [[|G_r(\theta_i)|f_i(t)][G_M(0)|f_d^*(t) + |G_M(\theta_i)|f_i^*(t)$$
$$+ W|G_r(\theta_i)|f_i^*(t)] \quad (1.27)$$
$$= |G_M(\theta_i)G_r(\theta_i)||f_i(t)|^2 + W|G_r(\theta_i)|^2|f_i(t)|^2$$
$$= |G_M(\theta_i)G_r(\theta_i)|P_i + W|G_r(\theta_i)|^2P_i$$

Then,

$$2[y_r(t)y_T^*(t)] = 2W|G_r(\theta_i)|^2 P_i + 2|G_M(\theta_i)G_r(\theta_i)|P_i \quad (1.28)$$

Substitution of (1.28) into (1.26) gives

$$dW/dt = 2[y_r(t)y_T^*(t)] \quad (1.29)$$

Equation (1.29) can be translated into a practical circuit configuration representing a feedback loop which could be subjected to an appropriate algorithm for automatic adjustment of the weighter parameters to remove the interference signal from the total output signal y_T.

Development of the Block Diagram of an Adaptive System

The right-hand side of (1.29) includes the product of $y_r(t)$, which is the output signal at the rth AUX port and $y_T^*(t)$, which is the complex conjugate of the combiner output signal or the total output signal $y_T(t)$.

In Figure 1.9, instead of delivering the signal $y_r(t)$, at the AUX-r output port, directly to the weighter, it is divided into two parts. One part is applied at one of the two input ports of a multiplier while the other part is delivered to the input port of the weighter. At the output of the weighter, the signal $y_r'(t)$ is combined with the main port signal $y_M(t)$ by a combiner Σ to give the total output signal y_T.

The signal $y_T(t)$ is now divided into two parts where one part will appear at the output port of the signal processing network as the final output signal $y_0(t)$ to be delivered to the radio receiver. The other part is applied at the second input port of the multiplier. The multiplier will convert $y_T(t)$ into its complex conjugate $y_T^*(t)$ and multiply with $y_r(t)$ to produce an output signal $y_r(t)y_T^*(t)$. This signal is now passed through a gain block of 2 to give $2[y_r(t)y_T^*(t)]$, which is the right-hand side of (1.29). This should now be equal to the time derivative of the weighter, dW/dt.

Now, let an integrator circuit be placed in front of the control port of the weighter W so that the weighter is connected to the output port of the integrator. Then, from the input port of the integrator circuit, looking into the integrator, one would see dW/dt; while looking toward the multiplier, one would see $2[y_r(t)y_T^*(t)]$ and, hence, (1.29) will be satisfied.

The entire operation can now be represented by the block diagram shown in Figure 1.9, which represents a basic single-loop adaptive array system.

Note that both configurations of the signal processing network, given in Figure 1.8 and Figure 1.9, would yield the same result. However, as mentioned earlier, the method in Figure 1.8 is tedious and time-consuming,

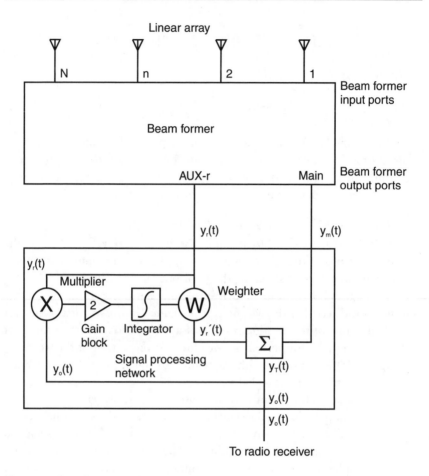

Figure 1.9 A basic single-loop adaptive array.

especially if more than one active AUX output port is involved. The configuration in Figure 1.9 is therefore more practical.

1.6 Adaptive Signal Processing

In adaptive signal processing, the cancellation of the undesired signals is carried out automatically with the help of a feedback circuit controlled by an appropriate algorithm.

The mechanism of adaptive signal processing can utilize either an analog or digital technique. In the past, an LMS algorithm has been used almost exclusively in conjunction with a feedback loop, or nulling loop as it is called, to remove the interference signals. However, in recent days, although the LMS algorithm is still widely used, the RLS algorithm is becoming increasingly popular in digital applications.

Many excellent books and papers are available that deal with different types of adaptive signal processing, an area which is beyond the scope of this book.

Reference

[1]   Kraus, J. D., *Antennas,* New York: McGraw-Hill, 1988, Secs. 11-7 and 12-2.

2

Simplified Analysis

Introduction

The function of an adaptive array is to remove or reduce any undesired or interference signal present in the signal received by a radio receiver. The degree of removal of the undesired signals from the total received signal is measured by a quantity called the *null depth*, which is defined as the ratio of the amount of the undesired signal contained in the signal received directly by a radio receiver without an adaptive array and that received through an active adaptive array.

In this chapter, expressions are derived to predict the null depth using the radiation pattern of the beam former being employed and the location of the interference signal source in question relative to the array boresight.

The analysis is carried out starting with the simplest model of an adaptive system having a single feedback or nulling loop in conjunction with a single interference signal source. Then the number of loops and the interference signal sources are increased step by step to the final configuration containing a large number of loops and interference signal sources. In addition, an attempt has been made for each section to be independent and self-sufficient.

At a first glance, the mathematics involved in the analysis may appear to be somewhat complicated. However, a closer look will reveal that all the derivations have been carried out using the four basic mathematical operations, namely, addition, subtraction, multiplication, and division.

2.1 The Signals

In an adaptive array system, a signal travels from the transmitting signal source to the signal processing network. Along its way, it passes through various circuit components, devices, and subsystems. In this section, the signal, as it appears at various stages along its path, is examined.

2.1.1 Signals at the Input and Output Ports of a Beam Former

Consider an $N \times N$ orthogonal beam former whose N input ports are connected to an N-element linear array. There is a desired signal source located along the array boresight and a number of interference signal sources randomly distributed on both sides of the boresight (Figure 2.1). It is assumed that none of the interference signal sources lies along any of the beams of the beam former. It is also assumed that all the signals, desired or undesired, are uncorrelated.

All elements of the linear array will intercept signals from all signal sources, desired or undesired, and deliver them to their respective input ports of the beam former. Hence, all the input ports of the beam former will carry signals from all signal sources. Again, all of these signals will be transported to the output ports of the beam former. However, since the desired signal

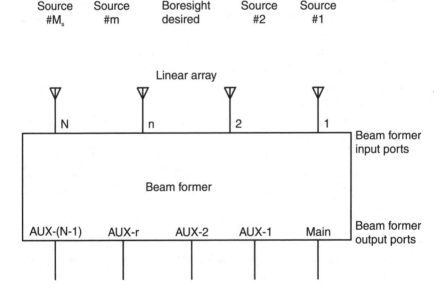

Figure 2.1 Beam former and the signals.

source is located along the boresight or the main beam, only the main output port will carry the desired signal while all the AUX ports, including the main port, will carry the signals from all the interference signal sources (Figure 2.2).

2.1.2 Space Transfer Functions

The signal transmitted by a signal source travels through space to reach a given array element. Consequently, the signal at the array element will be modified by a transfer function or gain function corresponding to the space between the signal source and the array element.

It has been assumed that the desired signal source is located along the array boresight. Therefore, the signals received by all array elements from the desired signal source will be in phase and hence the phase difference between any two adjacent array elements will be zero. On the other hand, all interference signal sources are located off the array boresight and hence the phase of the signal received by any given array element from any given interference signal source will be different from other array elements. As (1.4a) indicates, for a $\lambda/2$ spaced linear array, the phase of the signal received by the array element n, relative to the array element 1, from the mth signal source, located at $+\theta_m$ off the array boresight, is given by

$$\psi_{nm} = (n-1)\pi\sin\theta_m = (n-1)\psi_m \qquad n = 1, 2, \ldots, N \quad (2.1)$$

where $\psi_m = \pi\sin\theta_m$ is the phase difference between two adjacent array elements corresponding to the signal source m (1.4b). Let the transfer function or gain between the nth array element and the mth signal source be S_{nm} where S_{nm} is given by

$$S_{nm} = e^{-j\psi nm} \qquad n = 1, 2, \ldots, N \quad (2.2)$$

The transfer functions or gains between different input ports of the beam former and different interference signal sources could be represented by the values shown in Table 2.1 (Figure 2.2). Note that for the desired signal, with the source located along the array boresight, $\theta = 0$, and so the corresponding phase difference ψ_d between two adjacent array elements is also zero. Therefore, from (2.2), the transfer function S_{nd} between the nth array element and the desired signal source will be unity.

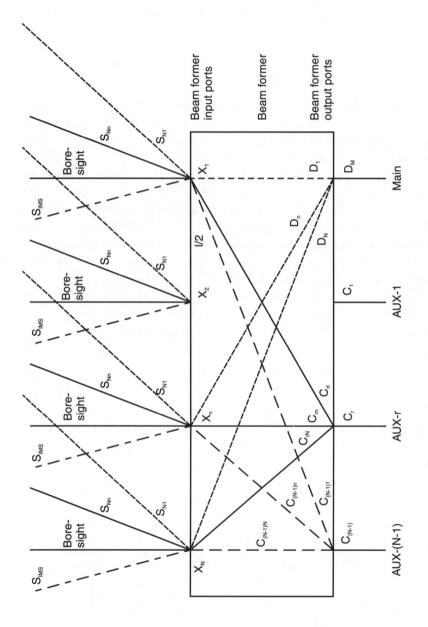

Figure 2.2 Free space and beam former transfer function signals at the input and output ports of the beam former.

Table 2.1

Beam Former	Interference Signal Sources		
Input Ports	1	2	*m*
1	S_{11}	S_{12}	S_{1m}
2	S_{21}	S_{22}	S_{2m}
n	S_{n1}	S_{n2}	S_{nm}
N	S_{N1}	S_{N2}	S_{Nm}

2.1.3 Signals at the Input Ports of the Beam Former

Let a_1, a_2, a_m be the amplitudes associated with the signals from the interference signal sources #1, #2, ... #m .... Then, the amplitude of the signal at the nth array element, due to the mth interference signal source, will be given by

$$X_{nm} = S_{nm} a_m \qquad (2.3a)$$

and that at the nth array element due to the desired signal source will be given by

$$X_{nd} = S_{nd} a_d = a_d \qquad (2.3b)$$

where a_d is the amplitude of the desired signal and $S_{nd} = 1$. Since the nth array element is directly connected to the nth input port of the beam former, both X_{nm} and X_{nd} will also appear at the nth input port of the beam former. Therefore, if X_n represents the sum of the amplitudes of the signals from all signal sources, desired and undesired, at the nth input port of the beam former, then

$$X_n = X_{n1} + X_{n2} + \ldots + X_{nm} + \ldots + X_{nd} \qquad (2.4)$$
$$= S_{n1} a_1 + S_{n2} a_2 + \ldots + S_{nm} a_m + \ldots + a_d$$

The sums of the amplitudes of the signals from all signal sources at different input ports of the beam former are shown in Table 2.2.

2.1.4 Beam Former Transfer Functions

When a signal at an input port of the beam former is transported to one of its output ports, it is further modified by the transfer function between the input and the output ports of the beam former.

Table 2.2

$$X_1 = X_{11} + X_{12} + \ldots + X_{1m} + \ldots + X_{1d} = S_{11}a_1 + S_{12}a_2 + \ldots + S_{1m}a_m + \ldots + a_d$$
$$X_2 = X_{21} + X_{22} + \ldots + X_{2m} + \ldots + X_{2d} = S_{21}a_1 + S_{22}a_2 + \ldots + S_{2m}a_m + \ldots + a_d$$
$$X_n = X_{n1} + X_{n2} + \ldots + X_{nm} + \ldots + X_{nd} = S_{n1}a_1 + S_{n2}a_2 + \ldots + S_{nm}a_m + \ldots + a_d$$
$$X_N = X_{N1} + X_{N2} + \ldots + X_{Nm} + \ldots + X_{Nd} = S_{N1}a_1 + S_{N2}a_2 + \ldots + S_{Nm}a_m + \ldots + a_d$$

The signal from a given input port of a beam former is transported to all of its output ports. Likewise, a given output port of a beam former contains signals from all of its input ports. However, during transportation from the input ports to the output ports, all signals are modified by the transfer functions between the input ports and the output ports (Figure 2.2).

Accordingly, the transfer functions or gains between the main output port of the beam former and all of its input ports will be designated as shown in Table 2.3. Similarly, the transfer functions or gains between all the AUX output ports of the beam former and all of its input ports will be designated as given in Table 2.4.

2.1.5 Signals at the Beam Former Output Ports

2.1.5.1 Signals at the Main Output Port

The signals at the main output port contain the desired signal and all interference signals. Hence, if D_n is the transfer function or gain between

Table 2.3

Beam former input ports	1	2	3	n	N
Transfer functions or gains	D_{M1}	D_{M2}	D_{M3}	D_{Mn}	D_{MN}

Table 2.4

	Beam Former Input Ports				
AUX Output Ports	**1**	**2**	**3**	***n***	***N***
1	C_{11}	C_{12}	C_{13}	C_{1n}	C_{1N}
2	C_{21}	C_{22}	C_{23}	C_{2n}	C_{2N}
3	C_{31}	C_{32}	C_{33}	C_{3n}	C_{3N}
r	C_{r1}	C_{r2}	C_{r3}	C_{rn}	C_{rN}
$N-1$	$C_{(N-1)1}$	$C_{(N-1)2}$	$C_{(N-1)3}$	$C_{(N-1)n}$	$C_{(N-1)N}$

the nth input port and the main output port, then the signal d_n at the main output port due to the signal X_n at the nth input port will be given by

$$d_n = Dn^*Xn \tag{2.5}$$

where Dn^* is the complex conjugate of D_n. Substitution of the value of Xn from (2.4) into (2.5) gives

$$\begin{aligned} d_n &= D_n^*(S_{n1}a_1 + S_{n2}a_2 + \ldots + S_{nm}a_m + \ldots + a_d) \\ &= D_n^*S_{n1}a_1 + D_n^*S_{n2}a_2 + \ldots + D_n^*S_{nm}a_m + \ldots + D_n^*a_d \tag{2.6} \\ &= G_{Mn1}a_1 + G_{Mn2}a_2 + \ldots + G_{Mnm}a_m + \ldots + G_{Mnd}a_d \end{aligned}$$

where G_{Mnm} is the transfer function or gain between the main output port and the mth signal source through the nth input port while G_{Mnd} is that between the main output port and the desired signal source through the nth input port.

Therefore, the total signal d_M at the main output port due to all signals at all input ports is given by

$$\begin{aligned} d_M = y_M &= d_1 + d_2 + \ldots + d_n + \ldots + d_N \tag{2.7} \\ &= D_1^*X_1 + D_2^*X_2 + \ldots + D_n^*X_n + D_N^*X_N \end{aligned}$$

Example 2.1

Using (2.6), express d_M in (2.7) in terms of transfer function or gain between the main output port and all the signal sources through all input ports of the beam former.

Answer

Using (2.6), d_M can be expressed as

$$\begin{aligned} d_M &= G_{M11}a_1 + G_{M12}a_2 + \ldots + G_{M1m}a_m + \ldots + G_{M1d}a_d \\ &\quad + G_{M21}a_1 + G_{M22}a_2 + \ldots + G_{M2m}a_m + \ldots + G_{M2d}a_d \\ &\quad + G_{Mn1}a_1 + G_{Mn2}a_2 + \ldots + G_{Mnm}a_m + \ldots + G_{Mnd}a_d \\ &\quad + G_{MN1}a_1 + G_{MN2}a_2 + \ldots + G_{MNm}a_m + \ldots + G_{MNd}a_d \\ &= (G_{M11} + G_{M21} + \ldots + G_{mn1} + \ldots + G_{MN1})a_1 \\ &\quad + (G_{M12} + G_{M22} + \ldots + G_{Mn2} + \ldots + G_{MN2})a_2 \\ &\quad + (G_{M1m} + G_{M2m} + \ldots + G_{Mnm} + \ldots + G_{MNm})a_m \\ &\qquad \ldots \qquad\quad \ldots \qquad\qquad \ldots \\ &\quad + (G_{M1d} + G_{M2d} + \ldots + G_{Mnd} + \ldots + G_{MNd})a_d \\ &= G_{M1}a_1 + G_{M2}a_2 + \ldots + G_{Mm}a_m + \ldots + G_{Md}a_d \tag{2.8a} \end{aligned}$$

where $G_{Mm} = G_{M1m} + G_{M2m} + \ldots + G_{Mnm} + \ldots + G_{MNm}$ (2.8b) is the transfer function or gain between the main output port and the mth signal source through all the input ports and

$$G_{Md} = G_{M1d} + G_{M2d} + \ldots + G_{Mnd} + \ldots + G_{MNd} \qquad (2.8c)$$

is the transfer function or gain between the main output port and the desired signal source through all the input ports.

2.1.5.2 Signals at the AUX Output Ports

The signal at a given AUX output port will contain signals from all the interference signal sources but none from the desired signal source. Hence, if C_{rn} is the transfer function or gain between the rth AUX output port and the nth input port, then the signal c_{rn} at the rth AUX output port due to the signal X_n at the nth input port is given by

$$c_{rn} = C_{rn}^* X_n \qquad (2.9)$$

where C_{rn}^* is the complex conjugate of C_{rn}. Substituting (2.6) in (2.9) gives

$$
\begin{aligned}
c_{rn} &= C_{rn}^*(S_{n1}a_1 + S_{n2}a_2 + \ldots + S_{nm}a_m + \ldots + a_d) \\
&= C_{rn}^* S_{n1}a_1 + C_{rn}^* S_{n2}a_2 + \ldots + C_{rn}^* S_{nm}a_m + \ldots + C_{rn}^* a_d \qquad (2.10) \\
&= G_{m1}a_1 + G_{m2}a_2 + \ldots + G_{mm}a_m + \ldots + G_{rnd}a_d
\end{aligned}
$$

where G_{rnm} is the transfer function or gain between the rth AUX output port and the mth signal source through the nth input port while G_{rnd} is that between the rth AUX output port and the desired signal source through the nth input port and is equal to zero since there is no desired signal at the rth AUX port.

Therefore, the total signal c_r at the rth AUX output port due to all the signals through all the input ports will be given by

$$
\begin{aligned}
c_r = y_r &= c_{r1} + c_{r2} + \ldots + c_{rn} + \ldots c_{rN} \qquad (2.11) \\
&= C_{r1}^* X_1 + C_{r2}^* X_2 + \ldots + C_{rn}^* X_n + \ldots + C_{rN} X_N
\end{aligned}
$$

The total signals at different AUX output ports are presented in Table 2.5.

Example 2.2

Using (2.10), express c_r in (2.11) in terms of the transfer function or gain between the rth AUX output port and all the signal sources through all the input ports of the beam former.

Table 2.5

$$c_1 = y_1 = C_{11}^*X_1 + C_{12}^*X_2 + \ldots + C_{1n}^*X_n + \ldots + C_{1N}X_N$$
$$c_2 = y_2 = C_{21}^*X_1 + C_{22}^*X_2 + \ldots + C_{2n}^*X_n + \ldots + C_{2N}X_N$$
$$c_3 = y_3 = C_{31}^*X_1 + C_{32}^*X_2 + \ldots + C_{3n}^*X_n + \ldots + C_{3N}X_N$$
$$c_r = y_r = C_{r1}^*X_1 + C_{r2}^*X_2 + \ldots + C_{rn}^*X_n + \ldots + C_{rN}X_N$$
$$c_{(N-1)} = y_{(N-1)} = C_{(N-1)1}^*X_1 + C_{(N-1)2}^*X_2 + C_{(N-1)n}^*X_n + \ldots + C_{(N-1)N}X_N$$

Answer

Keeping in mind that there is no desired signal at any of the AUX output ports, using (2.10), c_r can be expressed as

$$c_r = G_{r11}a_1 + G_{r12}a_2 + \ldots + G_{r1m}a_m + \ldots$$
$$+ G_{r21}a_1 + G_{r22}a_2 + \ldots + G_{r2m}a_m + \ldots$$
$$+ G_{rn1}a_1 + G_{rn2}a_2 + \ldots + G_{rnm}a_m + \ldots$$
$$+ G_{rN1}a_1 + G_{rN2}a_2 + \ldots + G_{rNm}a_m + \ldots$$
$$= (G_{r11} + G_{r21} + \ldots + G_{rn1} + \ldots + G_{rN1})a_1$$
$$+ (G_{r12} + G_{r22} + \ldots + G_{rn2} + \ldots + G_{rN2})a_2$$
$$+ (G_{r1m} + G_{r2m} + \ldots + G_{rnm} + \ldots + G_{rNm})a_m$$
$$\ldots$$
$$= G_{r1}a_1 + G_{r2}a_2 + \ldots + G_{rm}a_m + \ldots + G_{rMS}a_{MS} \quad (2.12a)$$

where $G_{rm} = G_{r1m} + G_{r2m} + \ldots + G_{rnm} + \ldots + G_{rNm}$ (2.12b)

is the transfer function or gain between the rth AUX output port and the mth signal source through all the input ports of the beam former.

2.1.6 Signals at the Input Ports of the Signal Processing Network

In general, the number of the input ports of the signal processing network is equal to the number of the output ports of the beam former where for every output port of the beam former there is a corresponding input port of the signal processing network. Accordingly, the main output port of the beam former is connected to the main input port of the signal processing network and all the AUX output ports of the beam former are connected to all the corresponding AUX input ports of the signal processing network. Therefore all signals appearing at the output ports of the beam former will be delivered to all the input ports of the signal processing network.

2.2 Two-Source, Two-Output System

Here, the system under discussion contains two transmitting signal sources while the associated beam former has two active output ports. The characteristics and performance of such a system are discussed in this section.

2.2.1 The Signals

The signal sources are arbitrarily distributed in free space. The transmitted signals will travel through space to reach the array location where they will be picked up by the array elements and delivered to the beam former. The output of the beam former will in turn be delivered to a signal processing network. The processed signal will then be applied at the input port of a radio receiver.

2.2.1.1 Signal Source and Location

A two-source, two-output system was shown in Figure 1.8, and also in Figure 1.9, where there is a desired signal source located along the array boresight and transmitting the desired signal $f_d(t)$. There is also an interference signal source located along some arbitrary direction $+\theta_i$ off the boresight and transmitting the interference signal $f_i(t)$. It is assumed that the source of the interference signal $f_i(t)$ is not located along any of the AUX beams of the beam former. It is also assumed that the signals $f_d(t)$ and $f_i(t)$ are uncorrelated.

2.2.1.2 Signals at the Beam Former Input Ports

There are two transmitting signal sources, the desired signal source and the interference signal source. Hence, the signals intercepted by an array element will contain both the desired and the interference signals. Each array element will deliver these intercepted signals to its corresponding input port of the beam former. Thus, each input port of the beam former will contain the desired signal $f_d(t)$ and the interference signal $f_i(t)$.

2.2.1.3 Transfer Functions

For the system configuration under consideration, the beam former has two output ports, the main output port and an arbitrary AUX-A output port. All other output ports of the beam former are match terminated. Thus, both the desired signal and the interference signal will appear at the main output port, while only the interference signal will appear at the AUX-A output port.

The transfer function or gain between the main output port and the desired signal source, located along $\theta = 0$, is designated by $|G_M(0)|$ and that between the main output port and the interference signal source, located along θ_i off the boresight, is designated by $|G_M(\theta_i)|$.

Also the transfer function or gain between the AUX-A output port and the interference signal source is designated by $|G_A(\theta_i)|$.

2.2.1.4 Signals at the Beam Former Output Ports

In terms of the gain functions, the signals $y_M(t)$ at the main output port and $y_A(t)$ at the AUX-A output port are given by

$$y_M(t) = |G_M(0)|f_d(t) + |G_M(\theta_i)|f_i(t) \qquad (2.13a)$$

$$y_A(t) = |G_A(\theta_i)|f_i(t) \qquad (2.13b)$$

Equations (2.13a) and (2.13b) represent the two signals appearing at the two active output ports of the beam former which will be delivered to the two corresponding input ports of the signal processing network.

2.2.2 Signal Processing Network

The signal processing network will receive signals from the two active output ports of the beam former, namely, the main port and the AUX-A port. It will modify the amplitude and phase of the AUX-A port signal, with the help of a weighter, such that when combined with the main port signal, it will remove the interference signal present in the main port signal. The resulting desired signal will then be applied to the connecting radio receiver.

2.2.2.1 Signal Processing Network with Manually Adjusted Weighter

As mentioned above, the signal processing network will receive only one AUX port signal from the beam former and hence will require only one weighter to process the incoming signals. Thus, since only one weighter is involved, a manually adjustable weighter could possibly be utilized.

Final Output Signals

Figure 2.3 shows the signal processing network employing a manually adjusted weighter W_A connected to the AUX-A output port. At the output of the weighter, the signal $y_A'(t)$ is given by

$$y_A'(t) = W_{AD}\, y_A(t) \qquad (2.14)$$

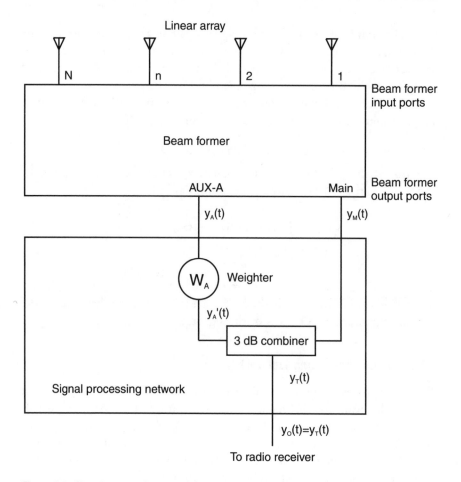

Figure 2.3 Signal processing network with manually adjusted weighter.

where the subscript D indicates that W_{AD} is a dimensionless quantity.

The signal $y_A'(t)$ is combined with the signal $y_M(t)$ from the main output port by a two-way 3-dB combiner to produce the total output signal $y_T(t)$ given by

$$y_T(t) = \frac{y_A'(t) + y_M(t)}{\sqrt{2}} = \frac{W_{AD}\, y_A(t) + y_M(t)}{\sqrt{2}} \qquad (2.15a)$$

or $\quad y_T(t) = \dfrac{W_{AD}\,|G_A(\theta_i)|\,f_i(t) + |G_M(0)|\,f_d(t) + |G_M(\theta_i)|\,f_i(t)}{\sqrt{2}}$

$$(2.15b)$$

This total output signal $y_T(t)$ will appear at the output port of the signal processing network as the final output signal $y_o(t)$, to be delivered at the input port of the radio receiver.

Optimum Weighter Value

From (2.15b) we can see that the interference signal $f_i(t)$ present in the total output signal $y_T(t)$ could be cancelled by adjusting W_{AD} so that

$$W_{AD}|G_A(\theta_i)|f_i(t) + |G_M(\theta_i)|f_i(t) = 0 \qquad (2.16)$$

$$\text{or} \qquad W_{AD} = -\left|\frac{G_M(\theta_i)}{G_A(\theta_i)}\right| \qquad (2.17)$$

Thus, W_{AD} in (2.17) is the optimum value of the weighter which would eliminate the interference signal present in the main output port signal $y_M(t)$ so that the final output signal $y_o(t)$, delivered to the radio receiver will be free of any interference signal.

As mentioned above, W_{AD} is a dimensionless quantity, normalized with respect to the weighter conversion factor m so that

$$W_{AD} = mW_A \qquad (2.18)$$

where W_A has the dimension of volt and m has the dimension of volt^{-1}. Hence, in (2.17), the optimum value of the weighter will become

$$W_{OP} = -\frac{1}{m}\left|\frac{G_M(\theta_i)}{G_A(\theta_i)}\right| \qquad (2.19)$$

2.2.2.2 Signal Processing Network with Self-Adjusting Weighter

The manual adjustment of the weighter, which involves the adjustment of both the amplitude and phase of the signal passing through the weighter, is obviously a difficult and slow process. It will be particularly difficult if there is more than one AUX output ports. Under these conditions, a self-adjusting weighter, controlled by a feedback loop, is appropriate.

Final Output Signal

A basic circuit, suitable to perform such an operation, was presented in Figure 1.9. A modified version of Figure 1.9 is shown in Figure 2.4.

The signal $y_A(t)$ at the AUX-A output port is divided into two equal parts by a two-way 3-dB splitter to give $[y_A(t)/\sqrt{2}]$. One part is delivered

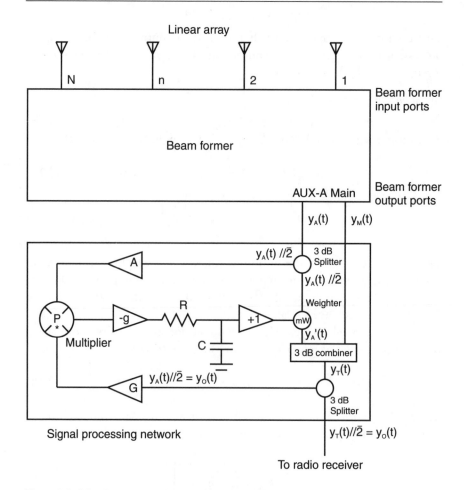

Figure 2.4 Adaptive array with a single feedback loop.

to the input port of a gain block with gain A while the other part is delivered to the input port of the weighter W_A. The output $[Ay_A(t)/\sqrt{2}]$ of the gain block A is applied at one of the two input ports of a multiplier while the output $y'_A(t)$ of the weighter W_A is applied at one of the two input ports of a two-way 3-dB combiner. The weighter output signal $y'_A(t)$, thus, is given by

$$y'_A(t) = \frac{mW_A\, y_A(t)}{\sqrt{2}} \tag{2.20}$$

As mentioned earlier, m is the conversion factor of the weighter with the dimension of volt^{-1}.

The main port signal $y_M(t)$ is applied to the second input port of the two-way combiner, which will combine $y_M(t)$ with $y_A'(t)$ to produce the total output signal $y_T(t)$ to appear at the output port of the combiner. The total output signal $y_T(t)$ is therefore given by

$$y_T(t) = \frac{y_A'(t) + y_M(t)}{\sqrt{2}} = \frac{1}{\sqrt{2}}\left[\frac{mW_A y_A(t)}{\sqrt{2}} + y_M(t)\right] \qquad (2.21)$$

The signal $y_T(t)$ is now divided into two equal parts by a two-way 3-dB splitter to give $[y_T(t)/\sqrt{2}]$. One part is applied to the second input port of the multiplier through a gain block with gain G while the other part will appear at the output port of the signal processing network as the final output signal $y_0(t)$ and will be delivered to the input port of the radio receiver.

The final output signal $y_0(t)$ is therefore given by

$$y_0(t) = \frac{y_T(t)}{\sqrt{2}} = \frac{1}{2}\left[\frac{mW_A y_A(t)}{\sqrt{2}} + y_M(t)\right] \qquad (2.22)$$

Steady-State Value of Self-Adjusting Weighter

As mentioned above, the multiplier receives signals $Ay_A(t)/\sqrt{2}$ at its one input port and $Gy_0(t)$ at its other input port. The multiplier first converts the signal $y_0(t)$ into its complex conjugate $y_0^*(t)$ and then performs the multiplication process. The resulting product signal at the output port of the multiplier will pass through another gain block with gain $-g$, an RC integrator circuit, and a unity gain stage and will finally be applied at the control input port of the weighter (Figure 2.4). Therefore, if p is the conversion factor of the multiplier with the dimension of volt^{-1}, then the signal $y_{mo}(t)$ at the output of the multiplier, after the gain block $-g$, will be given by

$$y_{mo}(t) = -GAgp\left[\frac{y_A(t)}{\sqrt{2}}\right]\left[\frac{y_T^*(t)}{\sqrt{2}}\right] \qquad (2.23)$$

$$= \frac{-\mu[W_A|y_A(t)|^2 - \sqrt{2}y_A(t)y_M^*(t)]}{m}$$

$$\text{where} \qquad \mu = \frac{GAgpm}{4} \qquad (2.24)$$

and represents the total gain factor associated with the feedback circuit controlling the weighter. As mentioned earlier, both p and m have the dimensions of volt^{-1} and hence μ has the dimension of power^{-1}.

Again, let $y_{ii}(t)$ be the signal at the input port of the integrator. Then, if RC = T, $y_{ii}(t)$ is given by

$$y_{ii}(t) = T(dW_A/dt) + W_A \qquad (2.25)$$

However, the signal at the output port of the gain block $-g$ is the same as that at the input port of the RC integrator circuit. Hence, (2.23) and (2.25) will give

$$\frac{dW_A}{dt} = -\frac{[1 + \mu|y_A(t)|^2]}{T}W_A - \frac{\sqrt{2}\mu[y_A(t)y_M^*(t)]}{mT} \qquad (2.26)$$

At steady-state $dW_{ASS}/dt = 0$. Hence,

$$m[1 + \mu|y_A(t)|^2]W_{ASS} + \sqrt{2}\mu[y_A(t)y_M^*(t)] = 0$$

$$\text{or} \qquad W_{ASS} = -\frac{\sqrt{2}}{m}\frac{\mu[y_A(t)y_M^*(t)]}{1 + \mu|y_A(t)|^2} \qquad (2.27)$$

Substituting (2.13) into (2.27) gives

$$W_{ASS} = -\frac{\sqrt{2}}{m}\frac{\mu p_i|G_A(\theta_i)G_M(\theta_i)|}{1 + \mu p_i|G_A(\theta_i)|^2} \qquad (2.28)$$

where $p_i = |f_i(t)|^2$ = the interference signal power. Note that if μp_i is very large, then

$$W_{ASS} = -\frac{\sqrt{2}}{m}\left|\frac{G_M(\theta_i)}{G_A(\theta_i)}\right| \qquad (2.29)$$

Equation (2.29) is identical to (2.17) representing the optimum weighter value which could completely remove the interference signal from the main output port signal. Thus, a self-adjusting weighter will adjust itself to the optimum value that can also completely remove the interference signal.

2.2.3 Final Output Power

The final output power p_o, appearing at the output port of the signal processing network, is given by

$$P_o = |y_o(t)|^2 = \frac{1}{4} \left[\frac{m W_A y_A(t)}{\sqrt{2}} + y_M(t) \right]^2 \tag{2.30}$$

$$= \frac{1}{4} \left[\frac{m^2 W_A^2 |y_A(t)|^2}{2} + |y_M(t)|^2 + \frac{2m}{\sqrt{2}} \text{Re}[W_A y_A(t) y_M(t)] \right]$$

Using (2.13),

$$|y_A(t)|^2 = |G_A(\theta_i)|^2 p_i \tag{2.31a}$$

$$|y_M(t)|^2 = |G_M(0)|^2 p_d + |G_M(\theta_i)|^2 p_i \tag{2.31b}$$

$$|y_A(t) y_M(t)| = |G_M(\theta_i) G_A(\theta_i)|^2 p_i \tag{2.31c}$$

$$p_i = |f_i(t)|^2 = \text{the interference signal power} \tag{2.31d}$$

$$p_d = |f_d(t)|^2 = \text{the desired signal power} \tag{2.31e}$$

Substitution of (2.31) into (2.30) gives

$$4p_{oSS} = \frac{m^2 W_A^2}{2} |G_A(\theta_i)|^2 p_i + |G_M(0)|^2 p_d + |G_M(\theta_i)|^2 p_i \tag{2.32}$$

$$+ \frac{2m W_A}{\sqrt{2}} |G_M(\theta_i) G_A(\theta_i)| p_i$$

At steady-state $W_A = W_{ASS}$. Substituting (2.13) and (2.27) into (2.30) and simplifying, the final steady-state output power, p_{oSS}, will be given by (Appendix A)

$$4p_{oSS} = p_{Md} + \frac{p_{Mi}}{[1 + \mu p_{Ai}]^2} \tag{2.33a}$$

$$= p_D + p_{\text{int}} \tag{2.33b}$$

where

$p_{Md} = |G_M(0)|^2 p_d$ = desired signal power at the main output port;

$p_{Mi} = |G_M(\theta_i)|^2 p_i$ = interference signal power at the main output port;

$p_{Ai} = |G_A(\theta_i)|^2 p_i$ = interference signal power at the rth AUX output port;

$p_D = p_{Md}$ = the desired signal power present in the total output power and

$$p_{int} = \frac{p_{Mi}}{[1 + \mu p_{Ai}]^2} \qquad (2.33c)$$

is the interference signal power present in the total output power. Equation (2.32) represents the final output power, appearing at the output port of the signal processing network to be delivered to the input port of the radio receiver.

2.2.4 Cancellation of Interference Signal and the Null Depth

Note that p_{int} given by (2.33c) represents the interference signal power contained in the final output power when the feedback loop is activated. When the feedback loop is deactivated, $\mu = 0$ and p_{int} becomes equal to p_{Mi}, which is the interference signal power contained in the main output port signal.

Now, since p_{int} ($\mu = 0$) is the total interference signal power with the feedback loop deactivated and p_{int} is that with the feedback loop activated, then the ratio of p_{int} ($\mu = 0$) and p_{int} is a measure of the degree of cancellation of the interference signal power achieved by the feedback loop. Let this ratio be represented by η; then,

$$\eta = \frac{p_{int}(\mu = 0)}{p_{int}} = \frac{p_{Mi}}{p_{int}} = [1 + \mu p_{Ai}]^2 \qquad (2.34a)$$

For $\mu p_{Ai} \gg 1$,

$$\eta = [\mu p_{Ai}]^2 \qquad (2.34b)$$

where η is called the *null depth* and the associated feedback loop is called the *nulling loop*.

Equation (2.34) indicates that larger null depth corresponds to larger μp_{Ai}. Therefore, in order to get a large null depth, large p_{Ai} and/or large μ is required.

However, p_{Ai} is the interference signal power appearing at the AUX-A output port and can be measured by a power meter. Hence, the power at all other AUX output ports could also be measured to determine the AUX port having the maximum output power which will correspond to the maximum null depth for a given feedback loop with a given value of μ. On the other hand, having chosen the suitable AUX output port, the value of μ required to achieve a desired null depth can also be determined.

Note that since μ has the dimension of power^{-1} and p_{Ai} has the dimension of power; μp_{Ai} is a dimensionless quantity and represents the total open loop gain of the feedback loop.

Note that the optimum weighter values for both manual and self-adjusting weighters are identical, and hence the value of η is expected to be about the same with both methods.

Example 2.3

Determine the value of the product μp_{Ai} required to obtain a desired null depth of 30 dB and hence present graphically the null depth η as a function of μp_{Ai}.

Answer

Given $\eta = 30$ dB = 1,000. Then, from (2.34a)

$$[1 + \mu p_{Ai}]^2 = 1,000$$

$$\text{or} \quad 1 + \mu p_{Ai} = 31.62$$

Therefore,

$$\mu p_{Ai} = 30.62 = 29.72 \text{ dB}$$

Note that use of (2.34b) will give $\mu p_{Ai} = 31.62 = 30$ dB. Therefore, the error introduced through the approximation is quite negligible.

Now, with the feedback loop disabled, $\mu p_{Ai} = 0$. Therefore, $\eta = 4 = 6$ dB. Similarly, for

$$\mu p_{Ai} = 35 \text{ dB} = 56.23$$

$$\eta = (57.23)^2 = 35 \text{ dB}$$

A plot of η as function of μp_{Ai} is shown in Figure 2.5.

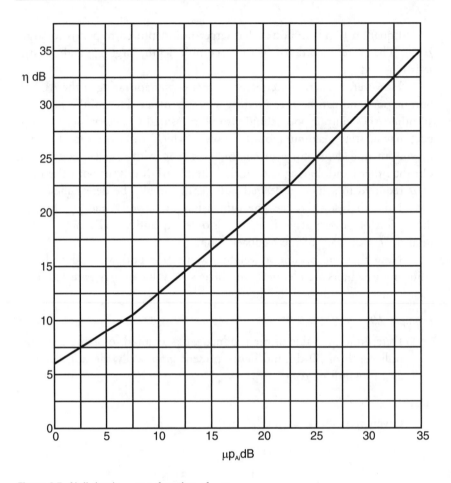

Figure 2.5 Null depth η as a function of μp_{Ai}.

From Figure 2.5 we can see that the degree of reduction of an interference signal depends primarily on the product μp_{Ai}. Therefore, to achieve higher reduction in the interference signal for a fixed AUX port output power p_{Ai}, the loop gain of the given feedback nulling loop will have to be increased. On the other hand, with a fixed loop gain μ for a given feedback nulling loop, the AUX port with the largest output power should be selected.

Example 2.4

An adaptive array system uses a 4×4 beam former as the feed network while the associated signal processing network contains a single feedback loop. There is an interference signal source located at $+20°$ off the array boresight. Select the AUX output port to give maximum null depth.

Answer

Let the total gain functions associated with the feedback loop be represented by μ. Then from (2.34), for a given value of μ, larger null depth requires larger p_{Ai}. Therefore, out of the three AUX output ports of the beam former, the one with largest output power should be selected.

Now, for a given output port AUX-r, the output power p_{ri} is

$$p_{ri} = |G_r(\theta_i)|^2 p_i$$

where p_i is the power of the interference signal. For $\theta = 20°$, the gain functions $|G_r(\theta_i)|$ for all the three AUX beams produced by a 4×4 beam former are obtained from Figure 1.6 as shown in Table 2.6. The output power p_{Ai} at AUX ports +1, +2, and −1 are also presented in Table 2.6.

From Table 2.6 it is seen that the output power at the AUX-(+1) port is the largest among the three output ports and hence will produce the maximum null depth.

Example 2.5

Assuming that AUX-(+1) output port in Example 2.4 provides a null depth of 30 dB, determine the expected null depths provided by the other two AUX output ports.

Answer

From Example 2.3, for a 30-dB of null depth,

$$\mu p_{(+1)i} = \mu p_i |G_{(+1)}(\theta_i)|^2 = 30.62$$

$$\text{but} \quad \mu p_{(+2)i} = \mu p_i |G_{(+2)}(\theta_i)|^2$$

$$\text{and} \quad \mu p_{(-1)i} = \mu p_i |G_{(-1)}(\theta_i)|^2$$

Table 2.6

AUX Port	+1	+2	−1
$|G_r(\theta_i)|$	0.8524	0.2437	0.2500
p_{Ai}	0.9194 p_i	0.0594 p_i	0.0625 p_i

Therefore,

$$\mu p_{(+2)i} = \frac{|G_{(+2)}(\theta_i)|^2}{|G_{(+1)}(\theta_i)|^2}[\mu p_{(+1)i}] = \frac{0.0594}{0.9194}(30.62) = 1.9783$$

or $\quad \eta_{+2} = 9.48$ dB

Again,

$$\mu p_{(-1)i} = \mu p_i \frac{|G_{(-1)}(\theta_i)|^2}{|G_{(+1)}(\theta_i)|^2}[\mu p_{(+1)i}] = \frac{0.0625}{0.9194}(30.62) = 2.0815$$

or $\quad \eta_{-1} = 9.78$ dB

2.3 Two-Source, Three-Output System

In this case, the system under consideration contains two transmitting signal sources as in the case with the system in Section 2.2. However, in the present system, the associated beam former has three active output ports.

2.3.1 The Signals

As before, signals transmitted by both the sources will be intercepted by the array and delivered to the beam former. But the present beam former has three output ports to deliver signals to three input ports of the signal processing network, which will process all three signals simultaneously to remove the interference signal before delivering it to the radio receiver.

2.3.1.1 Signal Source and Location

In this type of system configuration, there are two transmitting signal sources, a desired signal source and an interference signal source; and there are three active output ports, the main output port and two AUX output ports (Figure 2.6).

As in Section 2.2, the desired signal source, located along the array boresight, is transmitting the desired signal $f_d(t)$ while the interference signal source, located along an arbitrary direction $+\theta_i$ off the boresight, is transmitting the interference signal $f_i(t)$. It is assumed that $f_i(t)$ does not coincide with any of the AUX beams of the beam former. It is also assumed that $f_d(t)$ and $f_i(t)$ are uncorrelated.

All array elements will pick up signals from both sources and will deliver them to their respective input ports of the beam former. Signals from

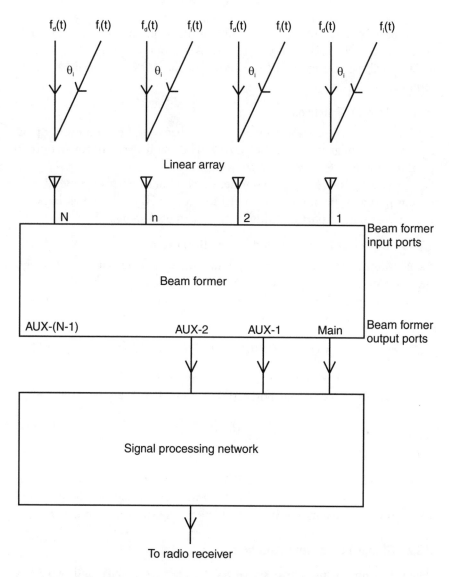

Figure 2.6 Typical system with two transmitting signal sources and three beam former output ports.

all the input ports of the beam former will subsequently be transported to all of its output ports.

Now, since the desired signal source is located along the array boresight or the main beam, the desired signal will appear only at the main output port. On the other hand, since the interference signal source does not coincide with either the main beam or any of the AUX beams, the interference signal will appear at every output port, including the main port. However, for the system under discussion, two arbitrary AUX output ports, AUX-1 and AUX-2, will be considered while the rest of the AUX ports will be match terminated.

2.3.1.2 Transfer Functions

Let $|G_M(0)|$ be the transfer function or gain between the main output port and the desired signal source, and $|G_M(\theta_i)|$ be that between the main output port and the interference signal source.

Again, let $|G_1(\theta_i)|$ be the transfer function or gain between the AUX-1 output port and the interference signal source and $|G_2(\theta_i)|$ be that between the AUX-2 output port and the interference signal source.

2.3.1.3 Signals at the Output Ports of the Beam Former

As mentioned above, the beam former has three active output ports, namely, the main output port and two AUX output ports, denoted by AUX-1 and AUX-2. As before [see (2.13a)], the signal $y_M(t)$ at the main output port of the beam former is given by

$$y_M(t) = |G_M(0)|f_d(t) + |G_M(\theta_i)|f_i(t) \tag{2.35}$$

However, there are now two AUX output ports. Let $y_1(t)$ and $y_2(t)$ be the signals at the output ports AUX-1 and AUX-2, respectively. Then,

$$y_1(t) = |G_1(\theta_i)|f_i(t) \tag{2.36a}$$

$$y_2(t) = |G_2(\theta_i)|f_i(t) \tag{2.36b}$$

Equations (2.35), (2.36a), and (2.36b) represent the three signals appearing at the three active output ports of the beam former and will be delivered to the three corresponding input ports of the signal processing network.

2.3.2 Signal Processing Network

The signal processing network receives the main port signal and two AUX port signals from the beam former. Hence, it will contain two feedback loops in association with two weighters to process the incoming signals in order to remove the interference signal.

2.3.2.1 Signal Processing Network with Manually Adjusted Weighters

Although adjustment of two weighters manually could be extremely difficult, in order to understand the mechanism involved, it is worth demonstrating. Hence, first the case of manually adjustable weighters is considered here.

Final Output Signal

The block diagram of a possible circuit is shown in Figure 2.7. The signals $y_1(t)$ and $y_2(t)$ from the output ports AUX-1 and AUX-2 are fed to weighters W_{1D} and W_{2D}, respectively, located inside the signal processing network. The weighted signals $y_1'(t)$ and $y_2'(t)$ are given by [see (2.14)]

$$y_1'(t) = W_{1D}\, y_1(t) = W_{1D}|G_1(\theta_i)|f_i(t) \qquad (2.37a)$$

$$y_2'(t) = W_{2D}\, y_2(t) = W_{2D}|G_2(\theta_i)|f_i(t) \qquad (2.37b)$$

Signals $y_1'(t)$ and $y_2'(t)$ are then combined with the main port output signal $y_M(t)$ by an N-way combiner whose unused ports, if any, are match terminated. Thus, to combine three input signals, a four-way combiner could be used with one input port match terminated. Hence, at the output port of the combiner, the total output signal $y_T(t)$ is given by

$$y_T(t) = \frac{[y_1'(t) + y_2'(t) + y_M(t)]}{\sqrt{N}}$$

$$= \frac{W_{1D}|G_1(\theta_i)|f_i(t) + W_{2D}|G_2(\theta_i)|f_i(t) + |G_M(0)|f_d(t) + |G_M(\theta_i)|f_i(t)}{\sqrt{N}}$$

$$(2.38)$$

This total output signal $y_T(t)$ will appear at the output port of the signal processing network as the final output signal $y_o(t)$ and will be delivered to the associated radio receiver.

Optimum Values of the Weighters

To remove the interference signal from the total output signal $y_T(t)$, all terms in the right hand side of (2.38) containing the factor $f_i(t)$ must vanish; that is,

$$W_{1D}|G_1(\theta_i)|f_i(t) + W_{2D}|G_2(\theta_i)|f_i(t) + |G_M(\theta_i)|f_i(t) = 0 \qquad (2.39)$$

$$\text{or} \quad W_{1D}|G_1(\theta_i)| + W_{2D}|G_2(\theta_i)| = -|G_M(\theta_i)|$$

Accordingly, both W_{1D} and W_{2D} are adjusted, while the total output signal $y_T(t)$ is observed on a monitor until $y_T(t)$ reaches its minimum value. At this stage, the interference signal contained in $y_T(t)$ will be removed and the weighters W_{1D} and W_{2D} will reach their optimum values.

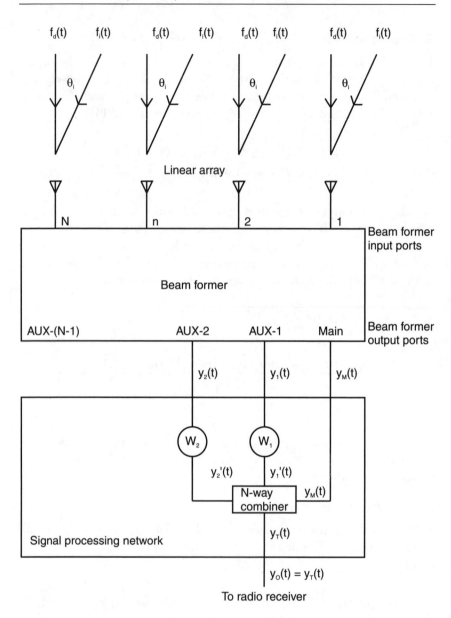

Figure 2.7 Two-source, three-output beam former with signal processing network having manually adjusted weighters.

2.3.2.2 Signal Processing Network with Self-Adjusting Weighters

Note, however, that simultaneous adjustment of two weighters manually to obtain a minimum value of $y_T(t)$ will be a very difficult process. On the other hand, self-adjusting weighters, controlled by feedback circuits, will automatically adjust themselves and therefore will be more practical.

Final Output Signal

Figure 2.8 shows a block diagram of a two-source, three-output system employing self-adjusting weighters. This is a modified version of Figure 2.7, where the weighters are automatically adjusted by feedback circuits. Each of the two AUX port output signals $y_1(t)$ and $y_2(t)$ is divided equally by a two-way divider to give $[y_1(t)/\sqrt{2}]$ and $[y_2(t)/\sqrt{2}]$ where one part of each is applied to the input port of a gain block with gain A and the other part is applied at the input port of the respective weighters W_1 and W_2. The output of the gain block is applied to one of the two input ports of a multiplier while the outputs of the two weighters, $y_1'(t)$ and $y_2'(t)$, are applied at the two input ports of an N-way combiner and are given by

$$y_1'(t) = \frac{mW_1 y_1(t)}{\sqrt{2}} = \frac{mW_1 |G_1(\theta_i)| f_i(t)}{\sqrt{2}} \tag{2.40a}$$

$$y_2'(t) = \frac{mW_2 y_2(t)}{\sqrt{2}} = \frac{mW_2 |G_2(\theta_i)| f_i(t)}{\sqrt{2}} \tag{2.40b}$$

where m is the weighter conversion factor having the dimension of $volt^{-1}$. The N-way combiner combines signals $y_1'(t)$ and $y_2'(t)$ with signal $y_M(t)$ from the main output port. (As before, a four-way combiner could be used where the fourth input port is match terminated.) The total output signal $y_T(t)$ at the combiner output port is therefore given by

$$y_T(t) = \frac{[y_1'(t) + y_2'(t) + y_M(t)]}{\sqrt{N}} \tag{2.41}$$

$$= \frac{1}{\sqrt{N}} \left[\frac{mW_1 y_1(t)}{\sqrt{2}} + \frac{mW_2 y_2(t)}{\sqrt{2}} + y_M(t) \right]$$

Signal $y_T(t)$ is further divided into two equal parts, to give

$$y_o(t) = \frac{y_T(t)}{\sqrt{2}} = \frac{1}{\sqrt{(2N)}} \left[\frac{mW_1 y_1(t)}{\sqrt{2}} + \frac{mW_2 y_2(t)}{\sqrt{2}} + y_M(t) \right] \tag{2.42}$$

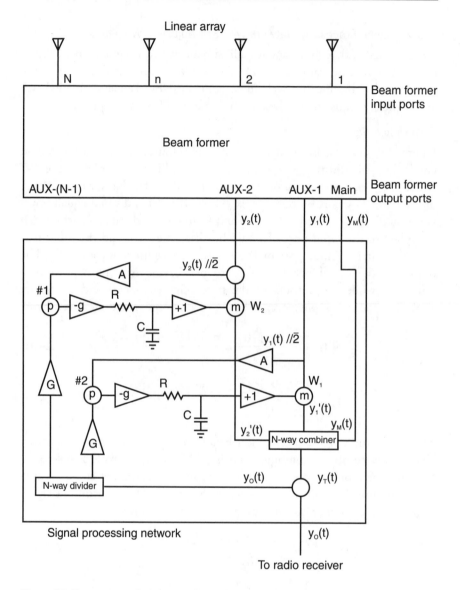

Figure 2.8 Two-source, three-output beam former with signal processing network having self-adjusting weighters.

where one part is applied at the input port of an N-way divider while the other part will appear at the output port of the signal processing network as the final output signal to be delivered to the input port of a radio receiver.

Steady-State Values of the Self-Adjusting Weighters

As mentioned above, one part of the total output signal $y_T(t)$ is applied at the input port of an N-way divider. The signal from each output port of this N-way divider is passed through a gain block with gain G and applied at the second input port of a multiplier which converts the signal to its complex conjugate and performs the multiplication process. The signal at the output port of a given multiplier is then passed through a gain block with gain $-g$, an RC integrator circuit, a unity gain block, and finally is delivered to the control port of the respective weighter W_1 or W_2. Thus, the output of the multiplier with signal $y_1(t)$ will be delivered to W_1 and that from $y_2(t)$ will be delivered to W_2 as shown in Figure 2.8.

Now, consider loop 1. Let p be the conversion factor for the multiplier having the dimension of volt^{-1}. Then according to (2.23), the signal $y_{om}(t)$ at the output port of the gain block $-g$ will be given by

$$y_{om}(t) = -g \frac{[Apy_1(t)]}{\sqrt{2}} \frac{[G y_o^*(t)]}{\sqrt{N}} = -\frac{gpAG[y_1(t)y_o^*(t)]}{\sqrt{(2N)}} \qquad (2.43)$$

$$= -\mu [W_1|y_1(t)|^2 + W_2 y_1(t)y_2^*(t)] - \frac{\sqrt{2}\mu [y_1(t)y_M^*(t)]}{m}$$

$$\text{where} \qquad \mu = \frac{GAgpm}{(2\sqrt{2})N} \qquad (2.44)$$

and represents the total gain factor associated with the feedback circuit controlling the weighters and since both p and m have the dimensions of volt^{-1}, μ has the dimension of power^{-1}.

Again, from (2.25), the signal $y_{ii}(t)$ at the input port of the integrator circuit RC is

$$y_{ii}(t) = T(dW_1/dt) + W_1 \qquad (2.45)$$

Equating the two equations, (2.43) and (2.45), gives

$$\frac{dW_1}{dt} = -\frac{[1 + \mu|y_1(t)|^2]W_1}{T} - \frac{[\mu y_1(t)y_2^*(t)]W_2}{T} - \frac{\sqrt{2}\mu[y_1(t)y_M^*(t)]}{mT}$$

$$(2.46a)$$

Similarly, loop 2 gives

$$\frac{dW_2}{dt} = -\frac{[1 + \mu|y_2(t)|^2]W_2}{T} - \frac{[\mu y_1^*(t)y_2(t)]W_1}{T} - \frac{\sqrt{2}\mu[y_2(t)y_M^*(t)]}{mT}$$

(2.46b)

At steady state $dW_1/dt = dW_2/dt = 0$ and hence (2.46) will become

$$[1 + \mu|y_1(t)|^2]W_{SS1} + [\mu y_1(t)y_2^*(t)]W_{SS2} + \frac{\sqrt{2}\mu[y_1(t)y_M^*(t)]}{m} = 0$$

(2.47a)

$$[1 + \mu|y_2(t)|^2]W_{SS2} + [\mu y_1^*(t)y_2(t)]W_{SS1} + \frac{\sqrt{2}\mu[y_2(t)y_M^*(t)]}{m} = 0$$

(2.47b)

Solving for W_{SS1} and W_{SS2}, the steady-state weighter values are given by

$$W_{SS1} = -\frac{\sqrt{2}\mu}{m} \frac{y_1(t)y_M^*(t)}{1 + \mu|y_1(t)|^2 + \mu|y_2(t)|^2}$$

(2.48a)

$$W_{SS2} = -\frac{\sqrt{2}\mu}{m} \frac{y_2(t)y_M^*(t)}{1 + \mu|y_1(t)|^2 + \mu|y_2(t)|^2}$$

(2.48b)

Substitution of (2.35) and (2.36) into (2.48) gives

$$W_{SS1} = -\frac{\sqrt{2}}{m} \frac{\mu|G_1(\theta_i)G_M(\theta_i)|p_1}{(1 + \mu p_{1i} + \mu p_{2i})}$$

(2.49a)

$$W_{SS2} = -\frac{\sqrt{2}}{m} \frac{\mu|G_2(\theta_i)G_M(\theta_i)|p_1}{(1 + \mu p_{1i} + \mu p_{2i})}$$

(2.49b)

where

$p_i = |f_i(t)|^2 =$ interference signal power   (2.50a)

$p_{1i} = |y_1(t)|^2 = |G_1(\theta_i)|^2 p_i =$ AUX-1 port output power   (2.50b)

$p_{2i} = |y_2(t)|^2 = |G_2(\theta_i)|^2 p_i =$ AUX-2 port output power   (2.50c)

2.3.3   Final Output Power

From (2.42), the final output power p_o will be given by

$$p_o = |y_o(t)|^2 = \frac{1}{2N}\left[\frac{m[W_1\,y_1(t)]}{\sqrt{2}} + \frac{m[W_2\,y_2(t)]}{\sqrt{2}} + y_M(t)\right]^2$$

$$(2.51)$$

and, hence, the final steady-state output power will be given by

$$p_{oSS} = \frac{1}{2N}\left[\frac{m[W_{SS1}\,y_1(t)]}{\sqrt{2}} + \frac{m[W_{SS2}\,y_2(t)]}{\sqrt{2}} + y_M(t)\right]^2 \quad (2.52)$$

Substituting (2.48) into (2.52) and simplifying give (Appendix B)

$$(2N)p_{oSS} = p_{Md} + \frac{p_{Mi}}{[1 + \mu p_{1i} + \mu p_{2i}]^2} \qquad (2.53a)$$

$$= P_D + P_{\text{int}} \qquad (2.53b)$$

where

$$p_D = p_{Md} = \text{desired signal power present in final output power} \qquad (2.54a)$$

and

$$p_{\text{int}} = \frac{p_{Mi}}{[1 + \mu p_{1i} + \mu p_{2i}]^2} \qquad (2.54b)$$

where p_{int} = interference signal power present in final output power;

$p_{Md} = |G_M(0)|^2 p_d$ = desired signal power at the main output port;

$p_{Mi} = |G_M(\theta_i)|^2 p_i$ = interference signal power at the main output port; and

$p_d = |f_d(t)|^2$ = power of the desired signal.

2.3.4   Null Depth

From (2.54b), p_{int} is the total interference signal power with the feedback loop engaged. With the feedback loop disabled (i.e., $\mu = 0$), the total interference signal power is p_{Mi}. Hence, the null depth η will be given by

$$\eta = \frac{p_{Mi}}{p_{\text{int}}} = [1 + \mu p_{1i} + \mu p_{2i}]^2$$

$$= [1 + \mu(p_{1i} + p_{2i})]^2 \qquad (2.55a)$$

$$\text{for } \mu(p_{1i} + p_{2i}) \gg 1, \ \eta = [\mu(p_{1i} + p_{2i})]^2 \qquad (2.55b)$$

From this, we can see that, for a given loop gain μ, the larger values of p_{1i} and p_{2i} will produce larger null depth. Since both p_{1i} and p_{2i} can be measured with a power meter, the two AUX output ports with highest output power levels could be selected.

Example 2.6

An adaptive array system uses a 4×4 beam former. It has two active feedback nulling loops. There is an interference signal source located at $+20°$ off the array boresight. Select the two AUX output ports to give the maximum null depth when simultaneously activated.

Answer

Using the data in Table 2.6 in Example 2.4, results for two simultaneous AUX ports, AUX-r and AUX-s, are presented in Table 2.7. When two AUX output ports are engaged simultaneously, ports +1 and −1 are expected to produce the maximum null depth.

Example 2.7

In Example 2.6, it is assumed that the two AUX output ports +1 and −1 jointly produce a null depth of 30 dB. Determine the contribution made by each port.

Answer

For a null depth of 30 dB,

$$[1 + \mu p_{(+1)i} + \mu p_{(-1)i}]^2 = 1,000$$

$$\text{or} \quad \mu p_{(+1)i} + \mu p_{(-1)i} = 30.62$$

Table 2.7

AUX Port	+1, +2	+2, −1	−1, +1
$\mu(p_{ri} + p_{si})$	$(0.9788)\,\mu p_i$	$(0.1219)\,\mu p_i$	$(0.9819)\,\mu p_i$

But, $\dfrac{\mu p_{(+1)i}}{\mu p_{(-1)i}} = \dfrac{\mu p_i |G_{+1}(\theta_i)|^2}{\mu p_i |G_{-1}(\theta_i)|^2} = \dfrac{0.9194}{0.0625} = 14.7104$

Therefore, $\mu p_{(-1)i} = \dfrac{30.62}{15.7104} = 1.9490$

Hence, $\mu p_{(+1)i} = 28.6710$

Therefore, the null depth produced by AUX port +1 alone will be

$$\eta_{+1} = [1 + 28.6710]^2 = 29.45 \text{ dB}$$

The contribution by the second port is therefore 0.55 dB.

From the above example, we can see that when only one interference signal source is involved, one nulling loop will probably be sufficient to obtain the maximum possible null depth. The contributions from extra loops appear to be negligible.

Example 2.8

An adaptive array system uses a 4×4 beam former as the feed network. There is an interference signal source located at $+40°$ off the array boresight. The system has two active feedback loops connected to AUX-(+1) and AUX-(+2) output ports because of the largest total gain (Figure 1.6). The loop gain μ has been so chosen that the total null depth is 30 dB. However, the source starts to move clockwise. Calculate the total null depth η obtained with both loops engaged. Also, calculate the null depth η_{+1} due to the primary loop (+1) alone and the contribution $\Delta\eta$ by the secondary loop (+2) until the source reaches the location where both AUX ports display the same output power.

Answer

From Figure 1.6, the radiation patterns of beams (+1) and (+2) cross over at $\theta = +48.6°$. Gains $|G_{+1}(\theta_i)|$ and $|G_{+2}(\theta_i)|$ between the interference signal source and the AUX-(+1) and AUX-(+2) output ports are calculated for different values of θ between $+40°$ and $+48.6°$ as shown in Table 2.8. The null depth η due to both loops and η_{+1} due to the primary loop (+1) alone are then calculated from the relation

Table 2.8

θ	$|G_{+1}(\theta_i)|$	$|G_{+2}(\theta_i)|$	$|G_{+1}(\theta_i)|^2$	$|G_{+2}(\theta_i)|^2$	η (dB)	η_{+1} (dB)	$\Delta\eta$ (dB)
40	0.8785	0.3672	0.7718	0.1348	30.00	28.65	1.35
42	0.8319	0.4398	0.6921	0.1934	29.80	27.74	2.06
44	0.7808	0.5093	0.6096	0.2594	29.64	26.68	2.96
46	0.7265	0.5750	0.5278	0.3306	29.54	25.50	4.04
48	0.6701	0.6362	0.4490	0.4047	29.49	24.17	5.32
48.6	0.6533	0.6533	0.4268	0.4268	29.49	23.76	5.73

$$\eta = [1 + \mu p_i (|G_{+1}(\theta_i)|^2 + |G_{+2}(\theta_i)|^2)]^2 \qquad (2.56a)$$

$$\eta_{+1} = [1 + \mu p_i (|G_{+1}(\theta_i)|^2)]^2 \qquad (2.56b)$$

The results, along with the contribution $\Delta\eta$ due to the secondary loop (+2), are also shown in Table 2.8.

Therefore, as the interference signal source moves clockwise away from beam (+1) toward beam (+2), the total null depth remains fairly the same. However, the contribution to the null depth by the primary loop (+1) decreases and that by the secondary loop (+2) increases as the crossover point of the two radiation patterns approaches, after which the role will reverse where loop (+2) will become the primary loop and loop (+1) will become the secondary loop.

Note that the maximum contribution by the secondary loop is about 6 dB.

Example 2.9

In Example 2.6, the AUX port pair (+1, −1) produces a null depth of 30 dB. Determine the expected null depths produced by (a) the pair (+1, +2) and (b) the pair (+2, −1).

Answer

From Example 2.7, a null depth of 30 dB corresponds to

$$\mu p_{(+1)i} = 28.6710$$
$$\mu p_{(-1)i} = 1.9490$$

Again, from Table 2.6,

$$p_{(+1)i} = 0.9194p_i$$
$$p_{(-1)i} = 0.0625p_i$$
Hence, $\mu p_i = 31.184$

Therefore, using Table 2.6,

(a) $\eta_{(+1,+2)} = [1 + 31.184(0.9194 + 0.0594)]^2 = 29.97$ dB

(b) $\eta_{(-1,+2)} = [1 + 31.184(0.0594 + 0.0625)]^2 = 13.63$ dB

Compare $\eta_{(+1,-1)} = [1 + 31.184(0.9194 + 0.0625)]^2 = 30.00$ dB

Thus, the two AUX port pairs (+1, +2) and (+1, −1) will produce practically the same result. This is because the gain at AUX-(+2) port is very close to the gain at AUX-(−1) port.

2.4 Two-Source, Multioutput System

The system under consideration in this section still has two transmitting signal sources. However, now the associated beam former has a large number of active output ports so that the signal processing network will have to process a large number of signals to remove the interference signal.

2.4.1 The Signals

Signals from both the transmitting sources, arbitrarily distributed in space, will travel through space to reach the array, which will pick up signals from both the sources. These signals will propagate through the beam former to all of its active output ports to be delivered to the signal processing network, which now will have to process a large number of signals to remove the interference signal.

2.4.1.1 Signal Source and Location

The system under consideration has two transmitting signal sources but a large number of active beam former output ports. One of the output ports is the main port; the rest are AUX ports (Figure 2.9). Note that the maximum number of active output ports is N, which is the total output ports of the beam former. Since one of the output ports is the main port, the maximum number of AUX output ports is $(N - 1)$.

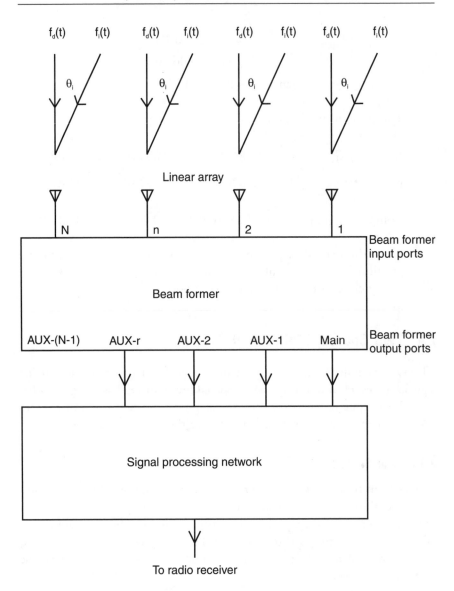

Figure 2.9 A typical system with two transmitting sources and multiple beam former output posts.

Out of the two transmitting signal sources, one is the desired signal source located along the array boresight or the main beam, transmitting the desired signal $f_d(t)$. The other is an interference signal source located along $+\theta_i$ off the boresight and transmitting the interference signal $f_i(t)$. As before, it is assumed that the interference signal source does not coincide with any of the AUX beams of the beam former and that signals $f_d(t)$ and $f_i(t)$ are uncorrelated.

As discussed earlier, all array elements will receive signals from both the desired and the undesired signal sources and, hence, all input ports of the beam former will carry signals from both signal sources.

Moreover, due to the locations of the two transmitting signal sources, the main output port will carry both the desired signal and the interference signal while all the AUX output ports will carry only the interference signal.

2.4.1.2 Transfer Functions

Let the transfer function or gain between the main output port and the desired signal source be denoted by $|G_M(0)|$ and that between the main output port and the interference signal source be denoted by $|G_{Mi}(\theta_i)|$.

Also, let the transfer function or gain between the rth AUX output port and the interference signal source be denoted by $|G_{ri}(\theta_i)|$ where $r = 1, 2, \ldots, (N-1)$.

2.4.1.3 Signals at the Beam Former Output Ports

Let $y_M(t)$ be the signal at the main output port and $y_{ri}(t)$ be that at the AUX-r output port. Then, $y_M(t)$ and $y_{ri}(t)$ are given by

$$y_M(t) = |G_M(0)| f_d(t) + |G_{Mi}(\theta_i)| f_i(t) \tag{2.57a}$$

$$y_{ri}(t) = |G_{ri}(\theta_i)| f_i(t) \qquad r = 1, 2, 3, \ldots \tag{2.57b}$$

Each output port of the beam former is connected to a corresponding input port of the signal processing network. Therefore, all signals from all the output ports, main and AUX, will be delivered to the signal processing network.

2.4.2 Signal Processing Network

As mentioned above, the signal processing network will receive a large number of signals from the beam former. It will now have to process those signals simultaneously to remove the interference so that a relatively pure desired signal could be delivered to the radio receiver.

2.4.2.1 Choice of Weighters

The signal processing network will receive a large number of signals from the beam former and hence will require an equally large number of weighters to process these signals. It will therefore be almost impossible to adjust all weighters simultaneously by hand. Use of automatic or self-adjusting weighters seems to be inevitable.

2.4.2.2 Signal Processing Network with Self-Adjusting Weighters

Figure 2.10 shows the block diagram of a signal processing network using self-adjustable weighters controlled by feedback circuits. Note that for clarity only the rth AUX output port has been shown in Figure 2.10.

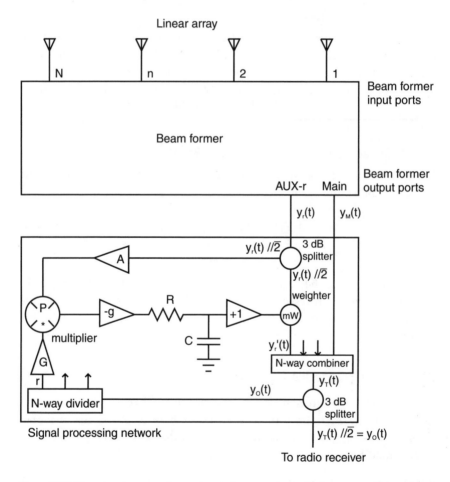

Figure 2.10 The signal processing network for a system with two transmitting signal sources and multiple beam former output ports. (Note that, for clarity, only the rth output port has been shown.)

Final Output Signal

As before, the signal $y_{ri}(t)$ from the rth AUX output port is divided into two equal parts $[y_{ri}(t)/\sqrt{2}]$ where one part is applied at the input port of a gain block with gain A while the other part is applied at the input port of the rth self-adjusting weighter W_r. The signal at the output port of the gain block A is applied at one of the two input ports of the rth multiplier while that at the output port of the weighter W_r is applied at one of the N input ports of an N-way combiner. At the output port of the weighter W_r, the signal $y'_{ri}(t)$ is given by

$$y'_{ri}(t) = \frac{mW_r y_{ri}(t)}{\sqrt{2}} = \frac{mW_r |G_{ri}(\theta_i)| f_i(t)}{\sqrt{2}} \qquad (2.58)$$

where m is the conversion factor of the weighter having the dimension of volt^{-1}. The N-way combiner combines $y'_{ri}(t)$ with the signal $y_M(t)$ from the main output port and the weighted signals from the output ports of all the other weighters. The total output signal $y_T(t)$ at the output port of the combiner is therefore given by

$$y_T(t) = \frac{[y_{1i}(t) + y_{2i}(t) + \ldots + y_{ri}(t) + \ldots + y_M(t)]}{\sqrt{N}} \qquad (2.59)$$

This total output signal $y_T(t)$ is further divided into two equal parts, where each part is equal to $y_o(t)$, given by

$$y_o(t) = \frac{y_T(t)}{\sqrt{2}} = \frac{[y_{1i}(t) + y_{2i}(t) + \ldots + y_{ri}(t) + \ldots + y_M(t)]}{\sqrt{(2N)}}$$

$$= \frac{1}{\sqrt{(2N)}} \left[\frac{mW_1 y_{1i}(t)}{\sqrt{2}} + \frac{mW_2 y_{2i}(t)}{\sqrt{2}} + \ldots + \frac{mW_r y_{ri}(t)}{\sqrt{2}} + \ldots + y_M(t) \right]$$

$$\qquad (2.60)$$

One part is applied at the input port of an N-way divider while the other part appears at the output port of the signal processing network as the final output signal and is delivered to the input port of the radio receiver.

Steady-State Values of the Self-Adjusting Weighters

As mentioned above, part of the total output signal $y_T(t)$ is applied at the input port of an N-way divider. The signal from the rth output port of the

N-way divider is passed through a gain block with gain G and is applied at the second input port of the rth multiplier, which will convert the signal into its complex conjugate before performing the multiplication process.

At the output port of the rth multiplier, the product signal is passed through a gain block $-g$, an RC integrator circuit, a unity gain stage, and is then applied to the control port of the rth weighter W_r.

The signal $y_{mo}(t)$, at the output of the multiplier, after the gain block $-g$, is therefore given by

$$y_{mo}(t) = -g \frac{[Apy_{ri}(t)]}{\sqrt{2}} \frac{[Gy_o^*(t)]}{\sqrt{N}} \tag{2.61}$$

$$= \frac{-gAGp}{\sqrt{(2N)}} [y_{1i}(t)y_o^*(t)]$$

where p is the conversion factor of the multiplier with the dimension of volt^{-1}. Substituting (2.60) into (2.61) and simplifying give

$$y_{mo}(t) = -\mu[W_1 y_{1i}^*(t)y_{ri}(t) + W_2 y_{2i}^*(t)y_{ri}(t) + \ldots$$

$$+ W_r|y_{ri}(t)|^2 + \ldots]$$

$$-\frac{\sqrt{2}\mu}{m} [y_M^*(t)y_{ri}(t)] \tag{2.62}$$

$$\text{where} \qquad \mu = \frac{gAGpm}{(2\sqrt{2})N} \tag{2.63}$$

and represents the total gain factor associated with the feedback circuit controlling the weighters. Again, with RC = T, the signal $y_{ii}(t)$ at the input port of the integrator circuit is given by

$$y_{ii}(t) = T(dW_r/dt) + W_r \tag{2.64}$$

Equating the two equations, (2.62) and (2.64), gives

$$\frac{dW_r}{dt} = -\frac{[1 + \mu|y_{ri}(t)|^2]W_r}{T} - \frac{\mu y_{ri}(t)}{T}[W_1 y_{1i}^*(t) + W_2 y_{2i}^*(t) + \ldots]$$

$$-\frac{\sqrt{2}\mu[y_{ri}(t)y_M^*(t)]}{mT} \tag{2.65a}$$

Similarly, for AUX-1 and AUX-2 output ports, the time derivatives of the corresponding weighters, dW_1/dt and dW_2/dt, are given by

$$\frac{dW_1}{dt} = -\frac{[1 + \mu|y_{1i}(t)|^2]W_1}{T} - \frac{\mu y_{1i}(t)}{T}[W_2 y_{2i}^*(t) + \ldots + W_r y_{ri}^*(t) + \ldots]$$

$$-\frac{\sqrt{2}\mu[y_{1i}(t)y_M^*(t)]}{mT} \qquad (2.65b)$$

$$\frac{dW_2}{dt} = -\frac{[1 + \mu|y_{2i}(t)|^2]W_2}{T} - \frac{\mu y_{2i}(t)}{T}[W_1 y_{1i}^*(t) + \ldots + W_r y_{ri}^*(t) + \ldots]$$

$$-\frac{\sqrt{2}\mu[y_{2i}(t)y_M^*(t)]}{mT} \qquad (2.65c)$$

At steady state $dW_1/dt = dW_2/dt = \ldots = dW_r/dt = \ldots = 0$. Therefore, (2.65) will give

$$[1 + \mu|y_{1i}(t)|^2]W_{SS1} + \mu y_{1i}(t)[y_{2i}^*(t)W_{SS2} + \ldots + y_{ri}(t)W_{SSr} + \ldots]$$

$$+ \frac{(\sqrt{2}\mu)}{m}[y_{1i}(t)y_M^*(t)] = 0 \qquad (2.66a)$$

$$[1 + \mu|y_{2i}(t)|^2]W_{SS2} + \mu y_{2i}(t)[y_{1i}^*(t)W_{SS1} + \ldots + y_{ri}^*(t)W_{SSr} + \ldots]$$

$$+ \frac{(\sqrt{2}\mu)}{m}[y_{2i}(t)y_M^*(t)] = 0 \qquad (2.66b)$$

$$[1 + \mu|y_{ri}(t)|^2]W_{SSr} + \mu y_{ri}(t)[y_{1i}^*(t)W_{SS1} + \ldots + y_{2i}^*(t)W_{SS2} + \ldots]$$

$$+ \frac{(\sqrt{2}\mu)}{m}[y_{ri}(t)y_M^*(t)] = 0 \qquad (2.66c)$$

From (2.66), we can see that to obtain the steady-state values of the weighters, we must solve a large number of simultaneous equations. A direct solution could be quite laborious and time-consuming. There is, however, the possibility of an indirect solution.

2.4.2.3 Indirect Solution for Steady-State Values of the Weighter

Equations (2.27) and (2.48) give the steady-state values of the weighters for a system with one interference signal source having two and three active output ports, respectively. Hence, by comparison, for a system with one interference signal source and a large number of active output ports, the steady-state value of the rth weighter will be given by (2.67).

$$W_{ASS} = -\frac{\sqrt{2}}{m}\frac{\mu[y_A(t)y_M^*(t)]}{1 + \mu|y_A(t)|^2} \tag{2.27}$$

$$W_{SS1i} = -\frac{\sqrt{2}\mu}{m}\frac{y_{1i}(t)y_M^*(t)}{1 + \mu|y_{1i}(t)|^2 + \mu|y_{2i}(t)|^2} \tag{2.48a}$$

$$W_{SS2i} = -\frac{\sqrt{2}\mu}{m}\frac{y_{2i}(t)y_M^*(t)}{1 + \mu|y_{1i}(t)|^2 + \mu|y_{2i}(t)|^2} \tag{2.48b}$$

$$W_{SSri} = \\ -\frac{\sqrt{2}\mu}{m}\frac{y_{ri}(t)y_M^*(t)}{[1 + \mu|y_{1i}(t)|^2 + \mu|y_{2i}(t)|^2 + \ldots + \mu|y_{ri}(t)|^2 + \ldots]} \tag{2.67}$$

where

$$|y_{1i}(t)|^2 = |G_1(\theta_i)||f_i(t)|^2 = p_{1i} = \text{AUX-1 output power}$$
$$\text{due to signal source } i;$$

$$|y_{2i}(t)|^2 = |G_2(\theta_i)||f_i(t)|^2 = p_{2i} = \text{AUX-2 output power} \tag{2.68}$$
$$\text{due to signal source } i;$$

$$|y_{ri}(t)|^2 = |G_r(\theta_i)||f_i(t)|^2 = p_{ri} = \text{AUX-}r \text{ output power}$$
$$\text{due to signal source } i.$$

Also, let

$$y_{1i}(t)y_M^*(t) = |G_1(\theta_i)G_M(\theta_i)||f_i(t)|^2 = q_{M1i};$$
$$y_{2i}(t)y_M^*(t) = |G_2(\theta_i)G_M(\theta_i)||f_i(t)|^2 = q_{M2i}; \tag{2.69}$$
$$y_{ri}(t)y_M^*(t) = |G_r(\theta_i)G_M(\theta_i)||f_i(t)|^2 = q_{Mri}.$$

where q_{Mri} represents the cross-gain interference signal power between the main and AUX-r output ports.

Then, (2.67) will give, for a large number of active AUX output ports,

$$W_{SS1i} = -\frac{\sqrt{2}\mu}{m} \frac{q_{M1i}}{1 + \mu[p_{1i} + p_{2i} + \ldots + p_{ri} + \ldots]}$$

$$W_{SS2i} = -\frac{\sqrt{2}\mu}{m} \frac{q_{M2i}}{1 + \mu[p_{1i} + p_{2i} + \ldots + p_{ri} + \ldots]} \qquad (2.70)$$

$$W_{SSri} = -\frac{\sqrt{2}\mu}{m} \frac{q_{Mri}}{1 + \mu[p_{1i} + p_{2i} + \ldots + p_{ri} + \ldots]}$$

2.4.2.4 Unified Expression for the Weighters

Let W_{SSi} represent the steady-state values of all weighters connected to all active AUX output ports and corresponding to the interference signal $f_i(t)$.

Similarly, let q_i represent the cross-gain interference signal powers between the main output port and all the active AUX output ports and corresponds to the interference signal $f_i(t)$.

Again, let λ_i represent the sum of the signal powers appearing at all active AUX outputs ports due to the interference signal $f_i(t)$, that is,

$$\lambda_i = p_{1i} + p_{2i} + \ldots + p_{ri} + \ldots \qquad (2.71)$$

Then all the expressions given by (2.70) could be represented by a single unified expression given by

$$W_{SSi} = -\frac{\sqrt{2}\mu}{m} \frac{q_i}{1 + \mu\lambda_i} \qquad (2.72a)$$

$$= -\sqrt{2}\mu' p \frac{q_i}{1 + \mu\lambda_i} \qquad (2.72b)$$

where $\quad \mu' = \mu/mp$

Note that since μ has the dimension of power^{-1} and both m and p have the dimension of volt^{-1}, μ' is a dimensionless quantity representing gain.

2.4.3 Final Output Power

The final output signal $y_o(t)$ is given by (2.60). Therefore, the final output power p_o will be given by

$$p_o = |y_o(t)|^2 = \frac{1}{4N}[mW_1 y_{1i}(t) + mW_2 y_{2i}(t) + \ldots \qquad (2.73)$$

$$+ mW_r y_{ri}(t) + \ldots + \sqrt{2}y_M(t)]^2$$

At steady state, the final output power is obtained by substituting (2.57) and (2.67) into (2.73). The calculations involved could be quite complicated. It therefore seems necessary to consider some alternate method to estimate the null depth.

2.4.4 Alternate Method for Estimating the Null Depth

From (2.71), for a system with one interference signal source i and one active AUX output port AUX-A, that is, a two-source two-output system,

$$\lambda_i = p_{Ai} \tag{2.74}$$

and hence, from (2.34), the null depth η will be given by

$$\eta = (1 + \mu p_{Ai})^2 = (1 + \mu \lambda_i)^2 \tag{2.75}$$

Again, for a system with one interference signal source i and two active AUX output ports, AUX-1 and AUX-2, that is, a two-source three-output system,

$$\lambda_i = p_{1i} + p_{2i} \tag{2.76}$$

So from (2.55), the corresponding null depth is

$$\eta = [1 + \mu(p_{1i} + p_{2i})]^2 \tag{2.77}$$
$$= (1 + \mu\lambda_i)^2$$

Similarly, for a system with one interference signal source i and r number of active AUX output ports, that is, a two-source multioutput system,

$$\lambda_i = p_{1i} + p_{2i} + \ldots + p_{ri} \tag{2.78}$$

Then, following (2.34) and (2.55), the corresponding null depth is given by

$$\eta = [1 + \mu(p_{1i} + p_{2i} + \ldots + p_{ri})]^2 \tag{2.79}$$
$$= (1 + \mu\lambda_i)^2$$

Therefore, if μ is known then by measuring the power at each AUX output port, the expected theoretical null depth could readily be estimated.

Therefore, the general expression for null depth due to an arbitrary interference signal source m, transmitting signal $f_m(t)$, is given by

$$\eta = (1 + \mu\lambda_m)^2 \tag{2.80}$$
$$= [1 + \mu(p_{1m} + p_{2m} + \ldots)]^2$$

2.5 Three-Source, Two-Output System

The system considered here involves three transmitting signal sources located arbitrarily in space. However, the associated beam former has only two active output ports so that the signal processing network will have to process only two input signals to remove the interference signal.

2.5.1 The Signals

Signals from all the three transmitting signal sources will be picked up by the array and will subsequently be delivered to the beam former. The beam former will transport these signals to its two active output ports from where they will enter the signal processing network, which, in turn, will remove the interference signals and deliver the desired signal to the radio receiver.

2.5.1.1 Signal Source and Location

The block diagram of a three-source, two-output system is shown in Figure 2.11. In this case, there are three transmitting signal sources: a desired signal source and two interference signal sources, 1 and 2. Also, only two of the beam former output ports are active. One of them is the main output port and the other is an AUX output port, designated the AUX-a port.

As before, the desired signal source is located along the array boresight and transmitting the desired signal $f_d(t)$. The two interference signal sources, 1 and 2, are located at $+\theta_1$ and $+\theta_2$ off the array boresight and transmitting the interference signals $f_1(t)$ and $f_2(t)$, respectively. It is assumed that none of the interference signal sources coincides with any of the AUX beams of the beam former. It is also assumed that all the three signals, $f_d(t)$, $f_1(t)$, and $f_2(t)$, are uncorrelated.

As mentioned earlier, all the array elements will pick up signals from all the three signal sources and deliver them to all input ports of the beam former, which in turn, will transport them to its output ports. However, due to the given locations of the signal sources, the desired signal will appear

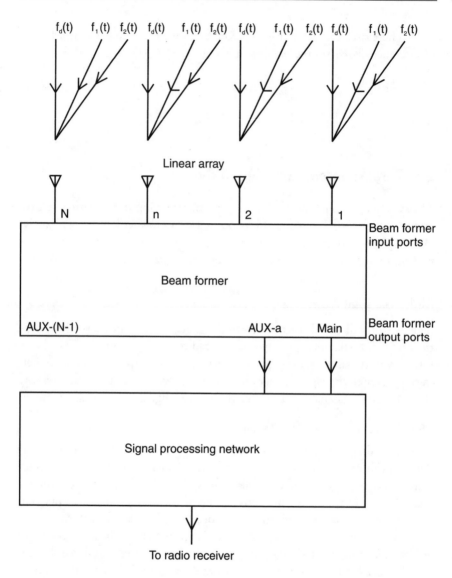

Figure 2.11 Typical system with three transmitting signal sources and two beam former output ports.

only at the main output port while signals from both the interference signal sources will appear at the main output port as well as at the AUX-a output port.

2.5.1.2 Transfer Functions

The transfer function between the main output port and the desired signal source will be designated $|G_M(0)|$ while those between the main output port and the interference signal sources 1 and 2 will be designated $|G_M(\theta_1)|$ and $|G_M(\theta_2)|$, respectively.

Also, the transfer functions between the AUX-a output port and the interference signal sources 1 and 2 will be designated by $|G_a(\theta_1)|$ and $|G_a(\theta_2)|$, respectively.

2.5.1.3 Signals at the Beam Former Output Ports

As mentioned above, the main output port will carry signals from all three signal sources. Hence, the signal $y_M(t)$ at the main output port is given by

$$y_M(t) = |G_M(0)|f_d(t) + |G_M(\theta_1)|f_1(t) + |G_M(\theta_2)|f_2(t) \qquad (2.81a)$$

Again, the AUX-a output port will carry only the two interference signals. Hence, the signal $y_a(t)$ at the AUX-a output port is given by

$$y_a(t) = |G_a(\theta_1)|f_1(t) + |G_a(\theta_2)|f_2(t) \qquad (2.81b)$$

Equations (2.81a) and (2.81b) represent the two signals appearing at the two active output ports of the beam former and will be delivered to the following stage, which is the signal processing network.

2.5.2 Signal Processing Network

As mentioned above, the beam former has only two active output ports. Since one of them is the main output port, it has only one AUX output port. Thus, the signal processing network will have only one feedback loop and hence only one weighter.

2.5.2.1 Choice of Weighters

Since the beam former contains only one weighter, either a manual or an automatic weighter can be used. Therefore, Figure 2.3 could be regarded as a possible representation of the signal processing network using a manually adjustable weighter while Figure 2.4 could be regarded as that using a self-adjusting weighter.

2.5.2.2 Signal Processing Network with Self-Adjusting Weighter

Although a manually adjusted weighter could be used in a single-loop system, automatic weighters are more commonly used. Hence, in the present case, a self-adjustable weighter, controlled by a feedback circuit, will be considered.

Final Output Signal

Considering Figure 2.4, the signal $y_a(t)$ is divided equally by a two-way divider to give $[y_a(t)/\sqrt{2}]$. One part is delivered to the input port of a gain block with gain A while the other part is applied at the input port of a self-adjusting weighter W_a. The output $[Ay_a(t)/\sqrt{2}]$ of the gain block A is applied at one of the two input ports of a multiplier while the output $y_a'(t)$ of the weighter W_a is applied at one of the two input ports of a two-way combiner and is given by

$$y_a'(t) = \frac{mW_a y_a(t)}{\sqrt{2}} = \frac{mW_a[\,|G_a(\theta_1)|f_1(t) + |G_a(\theta_2)|f_2(t)]}{\sqrt{2}}$$

(2.82)

where m is the conversion factor of the weighter having the dimension of volt^{-1}. The other input port of the two-way combiner receives signal $y_M(t)$ from the main output port. The combiner combines the two signals $y_a'(t)$ and $y_M(t)$ so that at the output of the combiner, the total signal $y_T(t)$ is given by

$$y_T(t) = \frac{y_a'(t) + y_M(t)}{\sqrt{2}} = \frac{1}{\sqrt{2}}\left[\frac{mW_a y_a(t)}{\sqrt{2}} + y_M(t)\right]$$

(2.83)

The total signal $y_T(t)$ is now divided into two equal parts by a two-way divider. One part is applied at the second input port of the multiplier through a gain block with gain G, while the other part appears at the output port of the signal processing network as the final output signal $y_0(t)$ to be delivered to the input port of a radio receiver. The final output signal $y_0(t)$ is therefore given by

$$y_0(t) = \frac{y_T(t)}{\sqrt{2}} = \frac{1}{2}\left[\frac{mW_a y_a(t)}{\sqrt{2}} + y_M(t)\right]$$

(2.84)

Steady-State Value of the Self-Adjusting Weighter

The multiplier receives signal $[Ay_a(t)/\sqrt{2}]$ at its one input port and the signal $Gy_0(t)$ at its other input port. It will convert $y_0(t)$ into its complex conjugate $y_0^*(t)$ and then perform the multiplication process.

At the output of the multiplier, the product signal is passed through a gain block with gain $-g$, an RC integrator circuit, and a unity gain stage before it is applied at the control input port of the weighter W_a. Hence, if p is the conversion factor of the multiplier having the dimension of volt^{-1}, then the signal $y_{om}(t)$ at the output of the multiplier, after the gain block $-g$, is given by

$$y_{om}(t) = -g\frac{[Apy_a(t)]}{\sqrt{2}}[Gy_o^*(t)] = -\frac{gpAG[y_a(t)y_o^*(t)]}{\sqrt{2}}$$

$$= -\mu W_a|y_a(t)|^2 - \frac{\sqrt{2}\mu[y_a(t)y_M^*(t)]}{m} \qquad (2.85)$$

$$\text{where} \qquad \mu = \frac{GAgpm}{4} \qquad (2.86)$$

and represents the total gain factor associated with the feedback circuit controlling the weighter and has the dimension of power^{-1} where both m and p have the dimension of volt^{-1}.

Again, from (2.45), the signal $y_{ii}(t)$ at the input port of the integrator circuit RC is given by

$$y_{ii}(t) = T\frac{dW_a}{dt} + W_a \qquad (2.87)$$

Equating the two equations, (2.85) and (2.87), gives

$$\frac{dW_a}{dt} = -\frac{[1 + \mu|y_a(t)|^2]W_a}{T} - \frac{\sqrt{2}\mu[y_a(t)y_M^*(t)]}{mT} \qquad (2.88)$$

At steady state $dW_a/dt = 0$ and hence (2.88) will give

$$[1 + \mu|y_a(t)|^2]W_{aSS} + \frac{\sqrt{2}\mu[y_a(t)y_M^*(t)]}{m} = 0 \qquad (2.89)$$

$$\text{or} \qquad W_{aSS} = -\frac{\sqrt{2}\mu[y_a(t)y_M^*(t)]}{m[1 + \mu|y_a(t)|^2]} \qquad (2.90)$$

Substitution of (2.81a) and (2.81b) into (2.90) gives

$$W_{aSS} = -\frac{\sqrt{2}\mu[|G_a(\theta_1)|f_1(t) + |G_a(\theta_2)|f_2(t)]\ [|G_M(0)|f_d(t) + |G_M(\theta_1)|f_1(t) + |G_M(\theta_2)|f_2(t)]}{1 + \mu[|G_a(\theta_1)|f_1(t) + |G_a(\theta_2)|f_2(t)]^2}$$

(2.91)

Keeping in mind that $f_d(t)$, $f_1(t)$ and $f_2(t)$ are uncorrelated, simplifying (2.91) gives the steady-state value of the weighter as

$$W_{aSS} = -\frac{\sqrt{2}\mu[|G_a(\theta_1)G_M(\theta_1)|p_1 + |G_a(\theta_2)G_M(\theta_2)|p_2]}{m[1 + \mu|G_a(\theta_1)|^2 p_1 + \mu|G_a(\theta_2)|^2 p_2]}$$

(2.92a)

$$= -\frac{\sqrt{2}\mu[|G_a(\theta_1)G_M(\theta_1)|p_1 + |G_a(\theta_2)G_M(\theta_2)|p_2]}{m[1 + \mu p_{a1} + \mu p_{b1}]}$$

where

$$p_1 = |f_1(t)|^2 = \text{the power of the interference signal } f_1(t)$$

$$p_2 = |f_2(t)|^2 = \text{the power of the interference signal } f_2(t)$$ (2.92b)

$$p_{a1} = |G_a(\theta_1)|^2 p_1 = \text{the power at AUX-a port due to signal } f_1(t)$$

$$p_{a2} = |G_a(\theta_2)|^2 p_2 = \text{the power at AUX-a port due to signal } f_2(t)$$

2.5.3 Final Output Power

Since the final output signal is $y_o(t)$, the final output power p_o will be given by

$$p_o = |y_o(t)|^2 = \frac{1}{4}\left[\frac{mW_a y_a(t)}{\sqrt{2}} + y_M(t)\right]^2$$

(2.93a)

Therefore, the final steady-state output power p_{oSS}, appearing at the output port of the signal processing network, is given by

$$p_{oSS} = \frac{1}{4}\left[\frac{m^2|y_a(t)|^2 W_{aSS}^2}{2} + |y_M(t)|^2 + \frac{2m}{\sqrt{2}}\text{Re}\{y_a(t)\, y_M(t)\, W_{aSS}\}\right]$$

(2.93b)

Substitution of (2.81) and (2.92) into (2.93b) and simplifying give (Appendix C)

$$4p_{oSS} = p_{Md} + p_{M1}\left[1 - \frac{\mu(p_{a1} + K_{a12}p_{a2})}{1 + \mu p_{a1} + mp_{a2}}\right]^2$$

$$+ p_{M2}\left[1 - \frac{\mu(p_{a2} + K_{a21}p_{a1})}{1 + \mu p_{a1} + \mu p_{a2}}\right]^2 \tag{2.94a}$$

$$= p_D + p_{int1} + p_{int2} \tag{2.94b}$$

where

$p_{Md} = |G_M(0)|^2 p_d$ = power of the desired signal at the main output port;

$p_{M1} = |G_M(\theta_1)|^2 p_1$ = power of the interference signal 1 at the main port;

$p_{M2} = |G_M(\theta_2)|^2 p_1$ = power of the interference signal 2 at the main port;

$p_d = |f_d(t)|^2$ = the desired signal power;

$p_D = p_{Md}$ = power of the desired signal contained in the final output power.

$$p_{int1} = p_{M1}\left[1 - \frac{\mu(p_{a1} + K_{a12}p_{a2})}{1 + \mu(p_{a1} + p_{a2})}\right]^2 \tag{2.95a}$$

$$p_{int2} = p_{M2}\left[1 - \frac{\mu(p_{a2} + K_{a21}p_{a1})}{1 + \mu(p_{1a} + p_{1b})}\right]^2 \tag{2.95b}$$

are the powers of the interference signals 1 and 2, respectively, contained in the final output power and where

$$K_{a12} = \frac{1}{K_{a21}} = \frac{|G_a(\theta_1)/G_a(\theta_2)|}{|G_M(\theta_a)/G_M(\theta_b)|} \tag{2.96}$$

Simplifying (2.94) gives

$$p_{int1} = p_{M1}\frac{[1 + \mu p_{a2}(1 - K_{a12})]^2}{[1 + \mu(p_{a1} + p_{a2})]^2} \tag{2.97a}$$

$$p_{int2} = p_{M2}\frac{[1 + \mu p_{a1}(1 - K_{a21})]^2}{[1 + \mu(p_{a1} + p_{a2})]^2} \tag{2.97b}$$

2.5.4 Null Depth

The theoretical null depths η_1 and η_2, corresponding to the interference signals $f_1(t)$ and $f_2(t)$, are therefore given by

$$\eta_1 = \frac{p_{M1}}{p_{int1}} = \frac{[1 + \mu(p_{a1} + p_{a2})]^2}{[1 + \mu p_{a2}(1 - K_{a12})]^2} \tag{2.98a}$$

$$\eta_2 = \frac{p_{M2}}{p_{int2}} = \frac{[1 + \mu(p_{a1} + p_{a2})]^2}{[1 + \mu p_{a1}(1 - K_{a21})]^2} \tag{2.98b}$$

For μp_{a1} and μp_{a2} much larger than one, (2.98) will reduce to

$$\eta_1 = \frac{[(p_{a1} + p_{a2})]^2}{[p_{a2}(1 - K_{a12})]^2} \tag{2.99a}$$

$$\eta_2 = \frac{[(p_{a1} + p_{a2})]^2}{[p_{a1}(1 - K_{a21})]^2} \tag{2.99b}$$

Example 2.10

An adaptive array system is built with a 4×4 beam former with the main beam along the array boresight and three AUX beams, +1, +2, and −1 located along +30°, ±90°, and −30°, respectively, as shown in Figure 1.6. There is a desired signal source located along the array boresight and two equal power interference signal sources, 1 and 2, located at +20° and +60°, respectively. Determine the AUX port which will produce the maximum null depths for both interference signals.

Answer

The desired AUX port will be designated the AUX-a port. Since both the interference signal sources are transmitting at the same power level, $p_1 = p_2$. Hence, using (2.92b), (2.97) gives

$$\eta_1 = \frac{[|G_a(\theta_1)|^2 + |G_a(\theta_2)|^2]^2}{[|G_a(\theta_2)|^2(1 - K_{a12})]^2} \tag{2.100a}$$

$$\eta_2 = \frac{[|G_a(\theta_1)|^2 + |G_a(\theta_2)|^2]^2}{[|G_a(\theta_2)|^2(1 - K_{a21})]^2} \tag{2.100b}$$

Now, from Figure 1.6, the gain functions between the interference signal sources 1 and 2 and the main, AUX-(+1), AUX-(+2) and

AUX-(−1) output ports are as given in Table 2.9. The resulting null depths are also shown in the table and indicate that the AUX-(+1) output port seems to be the best choice and will therefore be chosen as the desired AUX-a output port.

2.6 Three-Source, Three-Output System

In this case, the system under consideration contains three transmitting signal sources as discussed in Section 2.5. However, this time, the associated beam former has three active output ports and hence the signal processing network will have three input signals to process in order to eliminate the interference signals.

2.6.1 The Signals

Signals from all three signal sources will be intercepted by the linear array and delivered to the beam former, which, as before, transports them to its output terminals. The signals will then enter the signal processing network through the three active output ports of the beam former.

2.6.1.1 Signal Source and Location

Figure 2.12 shows the block diagram of a three-source, three-output system. Like the system in Section 2.5, this one also contains three transmitting signal sources. One of them is the desired signal source and the other two are the interference signal sources 1 and 2. However, in this case, the beam former has three active output ports. One of them is the main output port and the other two are AUX output ports, designated AUX-a and AUX-b ports.

Table 2.9

	Main	AUX-(+1)	Output Ports AUX-(+2)	AUX-(−1)
$G(\theta_1)$	0.4091	0.8524	0.2437	0.2500
$G(\theta_2)$	0.1907	0.3429	0.8926	0.2000
K_{a12}		1.1588	0.1273	0.5827
K_{a21}		0.8630	7.8574	1.7162
η_1 dB		33.10	1.80	15.76
η_2 dB		18.57	6.45	7.20

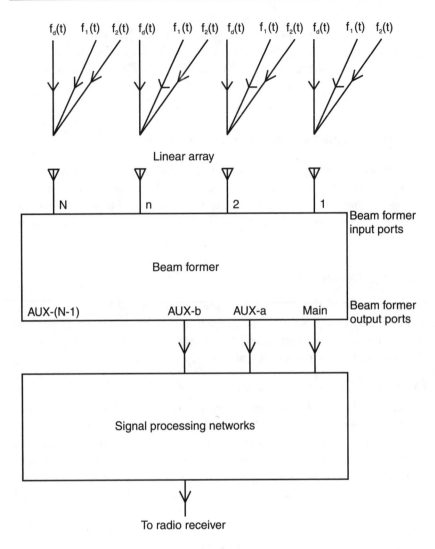

Figure 2.12 Typical system with three transmitting signal sources and three beam former output ports.

As before, the desired signal source is located along the array boresight and is transmitting the desired signal $f_d(t)$ while the two interference signal sources 1 and 2 are located along $+\theta_1$ and $+\theta_2$ off the boresight and are transmitting the interference signals $f_1(t)$ and $f_2(t)$, respectively. Also, none of the interference signal sources coincides with any of the AUX beams of the beam former, and all three signals are uncorrelated.

All array elements will then pick up signals from all three signal sources and will deliver them to all input ports of the beam former. At the output of the beam former, the main port will contain signals from all three signal sources while the two AUX output ports will carry only the signals from the two interference signal sources.

2.6.1.2 Transfer Functions

The transfer function between the main output port and the desired signal source will again be designated by $|G_M(0)|$ and those between the main output port and the interference signal sources 1 and 2 by $|G_M(\theta_1)|$ and $|G_M(\theta_2)|$, respectively.

Also, the transfer functions between the AUX-a output port and the two interference signal sources 1 and 2 will be designated by $|G_a(\theta_1)|$ and $|G_a(\theta_2)|$, respectively.

Likewise, the transfer functions between the AUX-b output port and the two interference signal sources 1 and 2 will be designated by $|G_b(\theta_1)|$ and $|G_b(\theta_2)|$, respectively.

2.6.1.3 Signals at the Beam Former Output Ports

Since the main output port will contain signals from all the three signal sources, the main port signal $y_M(t)$ will be given by

$$y_M(t) = |G_M(0)|f_d(t) + |G_M(\theta_1)|f_1(t) + |G_M(\theta_2)|f_2(t)$$
$$(2.101a)$$

Again, since both AUX-a and AUX-b output ports will carry signals from the two interference signal sources only, the output signals $y_a(t)$ at AUX-a port and $y_b(t)$ at AUX-b port will be given by

$$y_a(t) = |G_a(\theta_1)|f_1(t) + |G_a(\theta_2)|f_2(t) \qquad (2.101b)$$

$$y_b(t) = |G_b(\theta_1)|f_1(t) + |G_b(\theta_2)|f_2(t) \qquad (2.101c)$$

Equations (2.101) represent the three signals appearing at the three active output ports of the beam former and will therefore be delivered to the respective input ports of the signal processing network.

2.6.2 Signal Processing Network

The signal processing network will require three input ports to receive signals from the three active output ports of the beam former and hence will have one main input port and two AUX input ports. The two AUX input ports will be connected to two feedback loops in association with two weighters.

2.6.2.1 Choice of Weighters

As mentioned above, the signal processing network contains two weighters in association with two feedback loops. However, to tune two weighters simultaneously by hand to obtain the minimum total output power could prove to be a difficult and time-consuming process. It therefore seems more appropriate to use self-adjusting weighters instead.

2.6.2.2 Signal Processing Network with Self-Adjusting Weighters

As mentioned above, to avoid a tedious and time-consuming tuning process, for the system under discussion, self-adjusting weighters will be employed. Hence, Figure 2.8 will be used to represent the prospective signal processing network.

Final Output Signal

As discussed in an earlier section (see Figure 2.8), each of the two AUX output port signals, $y_a(t)$ and $y_b(t)$, is divided equally by a two-way divider to give $y_a(t)/\sqrt{2}$ and $y_b(t)/\sqrt{2}$. One part of each is applied at the input port of a gain block with gain A while the other part of each is applied at the input port of a weighter. The output of the gain block A is applied at one of the two input ports of a multiplier while the output of the weighter is applied at one of the N input ports of an N-way combiner.

The multipliers and weighters are designated according to the input signals. Thus, signal $y_a(t)$ will be applied to multiplier a and weighter a while signal $y_b(t)$ will be applied to multiplier b and weighter b.

The signals $y_a'(t)$ and $y_b'(t)$ at the output ports of the two weighters will be given by (2.40)

$$y_a'(t) = mW_a y_a(t) \qquad (2.102)$$

$$y_b'(t) = mW_b y_b(t) \qquad (2.103)$$

The signals $y_a'(t)$ and $y_b'(t)$ are then combined with the signal $y_M(t)$ from the main output port by the N-way combiner to produce the total output signal $y_T(t)$ given by (2.41)

$$y_T(t) = \frac{1}{\sqrt{N}} \left[\frac{mW_a y_a(t)}{\sqrt{2}} + \frac{mW_b y_b(t)}{\sqrt{2}} + y_M(t) \right] \qquad (2.104)$$

The total output signal $y_T(t)$ is now divided into two equal parts by a two-way divider where one part is applied at the input port of an N-way divider while the other part appears at the output port of the signal processing network as the final output signal $y_o(t)$ and is given by (2.42)

$$y_o(t) = \frac{y_T(t)}{\sqrt{2}} = \frac{1}{\sqrt{(2N)}} \left[\frac{mW_a y_a(t)}{\sqrt{2}} + \frac{mW_b y_b(t)}{\sqrt{2}} + y_M(t) \right]$$

(2.105)

Steady-State Values of the Self-Adjusting Weighters

Following the steps described in Section 2.3.2.2, the signal $y_{om}(t)$ at the output port of multiplier 1, after the gain block $-g$, is given by (2.43)

$$y_{om}(t) = -\mu [W_1 |y_1(t)|^2 + W_2 y_1(t) y_2^*(t)] - \frac{\sqrt{2}\mu [y_1(t) y_M^*(t)]}{m}$$

(2.106)

Similarly, the signal $y_{ii}(t)$ at the input port of the integrator circuit RC will be given by (2.45)

$$y_{ii}(t) = T(dW_1/dt) + W_1$$

(2.107)

Therefore, the time derivatives of the weighters for the two loops will be given by (2.46)

$$\frac{dW_a}{dt} = -\frac{[1 + \mu |y_a(t)|^2] W_a}{T} - \frac{[\mu y_a(t) y_b^*(t)] W_b}{T} - \frac{\sqrt{2}\mu [y_a(t) y_M^*(t)]}{mT}$$

(2.108a)

$$\frac{dW_b}{dt} = -\frac{[1 + \mu |y_b(t)|^2] W_b}{T} - \frac{[\mu y_a^*(t) y_b(t)] W_b}{T} - \frac{\sqrt{2}\mu [y_b(t) y_M^*(t)]}{mT}$$

(2.108b)

which at steady state will become

$$[1 + \mu |y_a(t)|^2] W_{aSS} + [\mu y_a(t) y_b^*(t)] W_{bSS} + \frac{\sqrt{2}\mu [y_a(t) y_M^*(t)]}{m} = 0$$

(2.109a)

$$[1 + \mu |y_b(t)|^2] W_{bSS} + [\mu y_a^*(t) y_b(t)] W_{bSS} + \frac{\sqrt{2}\mu [y_b(t) y_M^*(t)]}{m} = 0$$

(2.109b)

Solving for W_{aSS} and W_{bSS} gives

$$W_{aSS} = -\frac{\sqrt{2}\mu}{m} \frac{|y_a(t)y_M^*(t)|}{1 + \mu|y_a(t)|^2 + \mu|y_b(t)|^2} \tag{2.110a}$$

$$W_{bSS} = -\frac{\sqrt{2}\mu}{m} \frac{|y_b(t)y_M^*(t)|}{1 + \mu|y_a(t)|^2 + \mu|y_b(t)|^2} \tag{2.110b}$$

Substituting (2.101) into (2.110) and simplifying (keeping in mind that all the signals are uncorrelated) give

$$W_{aSS} = -\frac{\sqrt{2}\mu}{m} \frac{|G_a(\theta_1)G_M(\theta_1)|\,p_1 + |G_a(\theta_2)G_M(\theta_2)|\,p_2}{1 + \mu(p_{a1} + p_{a2}) + \mu(p_{b1} + p_{b2})} \tag{2.111a}$$

$$W_{bSS} = -\frac{\sqrt{2}\mu}{m} \frac{|G_b(\theta_1)G_M(\theta_1)|\,p_1 + |G_b(\theta_2)G_M(\theta_2)|\,p_2}{1 + \mu(p_{a1} + p_{a2}) + \mu(p_{b1} + p_{b2})} \tag{2.111b}$$

where

$p_1 = |f_1(t)|^2$ = power of interference signal 1;

$p_2 = |f_2(t)|^2$ = power of the interference signal 2;

$p_{a1} = |G_a(\theta_1)|^2 p_1$ = power of interference signal 1 at AUX-a output port;

$p_{a2} = |G_a(\theta_2)|^2 p_2$ = power of interference signal 2 at AUX-a output port;

$p_{b1} = |G_b(\theta_1)|^2 p_1$ = power of interference signal 1 at AUX-b output port;

$p_{b2} = |G_b(\theta_2)|^2 p_2$ = power of interference signal 2 at AUX-b output port;

and as before,

$$\mu = \frac{GAgpm}{(2\sqrt{2})N} \tag{2.112}$$

is the total gain factor associated with the feedback circuit controlling the weighter and has the dimension of power^{-1} while p and m are the conversion factors of the multiplier and weighter, respectively, and have the dimension of volt^{-1}.

2.6.3 Final Output Power

As mentioned above, the final output signal at the output port of the signal processing network is $y_o(t)$. Hence, the final output power will be given by

$$p_o = |y_o(t)|^2 = \frac{1}{2N}\left[\frac{mW_a\,y_a(t)}{\sqrt{2}} + \frac{mW_b\,y_b(t)}{\sqrt{2}} + y_M(t)\right]^2 \quad (2.113)$$

Therefore, at steady state, the final steady-state output power p_{oSS} will be given by

$$p_{oSS} = |y_o(t)|^2 = \frac{1}{2N}\left[\frac{mW_{aSS}\,y_a(t)}{\sqrt{2}} + \frac{mW_{bSS}\,y_b(t)}{\sqrt{2}} + y_M(t)\right]^2$$
$$(2.114)$$

Substituting (2.101) and (2.111) into (2.114) and simplifying, the resulting expression for the final output power is given by (Appendix D)

$$(2N)p_{oSS} = p_{Md} + p_{M1}\frac{[1 + \mu p_{a2}(1 - K_{a12}) + \mu p_{b2}(1 - K_{b12})]^2}{[1 + \mu(p_{a1} + p_{b1} + p_{a2} + p_{b2})]^2}$$

$$+ p_{M2}\frac{[1 + \mu p_{a1}(1 - K_{a21}) + \mu p_{b1}(1 - K_{b21})]^2}{[1 + \mu(p_{a1} + p_{b1} + p_{a2} + p_{b2})]^2} \quad (2.115a)$$

$$= p_D + p_{int1} + p_{int2} \quad (2.115b)$$

where

$p_{Md} = |G_M(0)|^2 p_d$ = power of the desired signal at the main output port;

$p_{M1} = |G_M(\theta_1)|^2 p_1$ = power of the interference signal 1 at the main port;

$p_{M2} = |G_M(\theta_2)|^2 p_2$ = power of the interference signal 2 at the main port;

$p_d = |f_d(t)|^2$ = the desired signal power;

$p_D = p_{Md}$ = power of the desired signal contained in the final output power.

$$p_{\text{int1}} = p_{M1} \frac{[1 + \mu p_{a2}(1 - K_{a12}) + \mu p_{b2}(1 - K_{b12})]^2}{[1 + \mu(p_{a1} + p_{b1} + p_{a2} + p_{b2})]^2} \qquad (2.116a)$$

$$p_{\text{int2}} = p_{M2} \frac{[1 + \mu p_{a1}(1 - K_{a21}) + \mu p_{b1}(1 - K_{b21})]^2}{[1 + \mu(p_{a1} + p_{b1} + p_{a2} + p_{b2})]^2} \qquad (2.116b)$$

are the interference signal powers contained in the final output power due to the interference signal sources 1 and 2, respectively, and where

$$K_{a12} = \frac{1}{K_{a21}} = \frac{|G_a(\theta_1)/G_a(\theta_2)|}{|G_M(\theta_1)/G_M(\theta_2)|} \qquad (2.117a)$$

$$K_{b12} = \frac{1}{K_{b21}} = \frac{|G_b(\theta_1)/G_b(\theta_2)|}{|G_M(\theta_1)/G_M(\theta_2)|} \qquad (2.117b)$$

2.6.4 Null Depth

The theoretical null depths η_1 and η_2, corresponding to the interference signals $f_1(t)$ and $f_2(t)$, respectively, are therefore given by

$$\eta_1 = \frac{p_{M1}}{p_{\text{int1}}} = \frac{[1 + \mu(p_{a1} + p_{b1} + p_{a2} + p_{b2})]^2}{[1 + \mu p_{a2}(1 - K_{a12}) + \mu p_{b2}(1 - K_{b12})]^2} \qquad (2.118a)$$

$$\eta_2 = \frac{p_{M2}}{p_{\text{int2}}} = \frac{[1 + \mu(p_{a1} + p_{b1} + p_{a2} + p_{b2})]^2}{[1 + \mu p_{a1}(1 - K_{a21}) + \mu p_{b1}(1 - K_{b21})]^2} \qquad (2.118b)$$

If the product of μ and the sum of all the powers, given in the numerators of (2.118), is much larger than unity, then (2.118) will become

$$\eta_1 = \frac{[p_{a1} + p_{b1} + p_{a2} + p_{b2}]^2}{[p_{a2}(1 - K_{a12}) + p_{b2}(1 - K_{b12})]^2} \qquad (2.119a)$$

$$\eta_2 = \frac{[p_{a1} + p_{b1} + p_{a2} + p_{b2}]^2}{[p_{a1}(1 - K_{a21}) + p_{b1}(1 - K_{b21})]^2} \qquad (2.119b)$$

Example 2.11

Consider an adaptive system utilizing a 4 × 4 beam former. There is a desired signal source along the boresight and two interference signal sources a and b, of equal strength, located along +20° and +60° off the boresight, respectively. Determine the two AUX output ports to give the maximum null depths for both interference signals.

Answer

Since both the interference signals have the same strength, (2.116) could be expressed in terms of the respective gain functions as

$$\eta_1 = \frac{[|G_a(\theta_1)|^2 + |G_a(\theta_2)|^2 + |G_b(\theta_1)|^2 + |G_b(\theta_2)|^2]^2}{[|G_a(\theta_2)|^2(1 - K_{a12}) + [|G_b(\theta_2)|^2(1 - K_{b12})]^2} \quad (2.120a)$$

$$\eta_2 = \frac{[|G_a(\theta_1)|^2 + |G_a(\theta_2)|^2 + |G_b(\theta_1)|^2 + |G_b(\theta_2)|^2]^2}{[|G_a(\theta_1)|^2(1 - K_{a21}) + [|G_b(\theta_1)|^2(1 - K_{b21})]^2} \quad (2.120b)$$

The required parameters are obtained from Figure 1.6 and presented in Table 2.10. Substitution of these parameters in (2.117) gives the null depths for a given pair of AUX output ports as shown in Table 2.11. Therefore, the AUX output pair $(-1, +1)$ seems to be the best choice.

Table 2.9, presented in Example 2.10, shows that two interference signals a and b could be reduced by a single loop with the resulting depths of about 33 and 19 dB, respectively, using the AUX-(+1) output port.

Table 2.10

	Main	AUX-(+1)	Output Ports AUX-(+2)	AUX-(−1)
$G_M(\theta_1)$	0.4091			
$G_M(\theta_2)$	0.1907			
$G_r(\theta_1)$		0.8524	0.2437	0.2500
$G_r(\theta_2)$		0.3429	0.8926	0.2000
K_{r12}		1.1588	0.1273	0.5827
K_{r21}		0.8630	7.8574	1.7162

Table 2.11

	AUX-(+1), (+2)	AUX-(+2), (−1)	AUX-(−1), (+1)
η_1 (dB)	8.0	3.0	54
η_2 (dB)	15.0	7.0	25

However as can be seen from Table 2.11, when two loops, in conjunction with the AUX-(+1) and AUX-(−1) output ports, are used, the null depth for interference signal a increases considerably, but that for signal b increases by about 6 dB.

Therefore, depending on the minimum tolerable null depth, one could perhaps get away with a single loop to null two interference signal sources. However, in situations where every dB of null depth counts, two nulling loops will probably be unavoidable.

2.7 Multisource, Two-Output System

Although in Section 2.4, a beam former with a large number of active output ports was considered, so far, the number of transmitting signal sources has been kept to a maximum of three. In this section, a system with a large number of transmitting signal sources is considered. However, the associated beam former will have only two active output ports.

2.7.1 The Signals

The elements of the linear array will pick up signals from all transmitting signal sources and deliver them to all input ports of the associated beam former, which, in turn, will transport them to all of its output terminals. However, since the beam former has only two active output ports, all of these signals will enter the signal processing network through these two output ports.

2.7.1.1 Signal Source and Location

In this case, the system involves a large number of transmitting signal sources. One of them is the desired signal source while the rest are interference signal sources scattered randomly on both sides of the desired signal source. However, for simplicity, only two of the beam former output ports will be regarded as active, while the rest will be match terminated. One of the active output ports will be the main port, while the other will be an AUX port, designated AUX-A port (Figure 2.13).

The desired signal source is located along the array boresight transmitting the desired signal $f_d(t)$ while the interference signal sources are located at $\theta_1, \theta_2, \ldots \theta_m, \ldots$, off the array boresight and transmitting interference signals $f_1(t), f_2(t), \ldots, f_m(t), \ldots$, respectively. It is assumed that none of these interference sources is located along any of the AUX beams of the beam former. It is also assumed that all signals, desired and undesired, are uncorrelated.

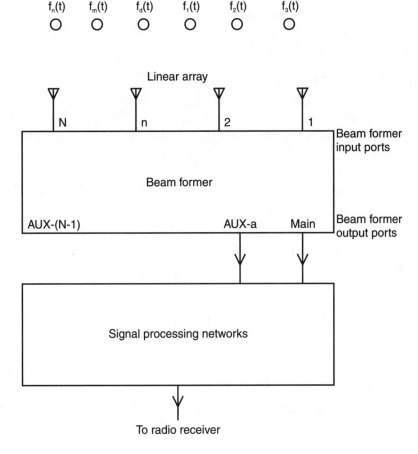

Figure 2.13 Adaptive system with multiple transmitting signal sources and two active beam former output ports.

All array elements will pick up signals from all sources, desired or undesired, and deliver them to all input ports of the beam former. However, since the desired signal source is located along the array boresight or the main beam of the beam former, the desired signal will appear only at the main output port. On the other hand, all the interference signals will appear at the main output port as well as at all AUX output ports.

2.7.1.2 Transfer Functions

The transfer function or gain between the main output port and the desired signal source is designated $|G_M(0)|$ while those between the main output port and the interference signal sources are designated $|G_M(\theta_1)|$, $|G_M(\theta_2)|$, $|G_M(\theta_3)|$, ..., $|G_M(\theta_m)|$, ...

Similarly, the transfer functions or gains between the AUX-A output port and the interference signal sources are designated by $|G_A(\theta_1)|, |G_A(\theta_2)|,$ $|G_A(\theta_3)|, \ldots, |G_A(\theta_m)|, \ldots$.

2.7.1.3 Signals at the Beam Former Output Ports

As before, signal $y_M(t)$ at the main output port will contain the desired signal and all interference signals. On the other hand, signal $y_A(t)$ at the AUX-A output port will carry only the interference signals. Therefore, the two output signals $y_M(t)$ and $y_A(t)$ are given by

$$y_M(t) = |G_M(\theta_1)|f_1(t) + |G_M(\theta_2)|f_2(t) + \ldots \qquad (2.121a)$$
$$+ |G_M(\theta_m)|f_m(t) + \ldots + |G_M(0)|f_d(t)$$

$$y_A(t) = |G_A(\theta_1)|f_1(t) + |G_A(\theta_2)|f_2(t) + \ldots \qquad (2.121b)$$
$$+ |G_A(\theta_m)|f_m(t) + \ldots$$

This equation represents the two signals appearing at the two active output ports of the beam former and will be delivered to the signal processing network.

2.7.2 Signal Processing Network

Although the number of signals involved is quite large, as mentioned above, the beam former has only two active output ports and, therefore, the signal processing network will require only two input ports to receive them from the beam former. Also, since one of the input ports is the main port, the signal processing network will have only one AUX input port and hence only one feedback loop to process the received signals.

2.7.2.1 Choice of Weighters

As mentioned above, the signal processing network has only one feedback loop and hence only one associated weighter. Hence, it will probably not be that difficult to tune a single weighter to obtain the minimum output power. However, due to the presence of the large number of interference signals, we may be required to increase the number of active AUX output ports, and hence the number of feedback loops, and thereby the number of associated weighters. It is therefore more appropriate to use a self-adjusting weighter with a single feedback loop which could later be increased to any desired number if necessary.

2.7.2.2 Signal Processing Network with Self-Adjusting Weighters

Following the discussions in Section 2.7.2.1, for the system under discussion, a self-adjusting weighter is used, for which the corresponding block diagram is shown in Figure 2.4.

Final Output Signal

Referring back to Figure 2.4, the signal $y_A(t)$ at the AUX-A output port is divided into two equal parts, $y_A(t)/\sqrt{2}$ whose one part is applied at the input port of a gain block with gain A while the other part is applied at the input port of a self-adjusting weighter W_A. The signal $[Ay_A(t)/\sqrt{2}]$ at the output of the gain block A is applied at one of the two input ports of a multiplier while the signal $y'_A(t)$ at the output of the weighter W_A is applied at one of the two input ports of a two-way combiner.

The weighted signal $y'_A(t)$ is then given by

$$y'_A(t) = \frac{mW_A y_A(t)}{\sqrt{2}} = \frac{mW_A[|G_A(\theta_1)|f_1(t) + |G_A(\theta_2)|f_2(t) + \ldots]}{\sqrt{2}}$$

(2.122)

As before, m is the conversion factor of the weighter having the dimension of volt^{-1}. The other input port of the combiner will receive the signal $y_M(t)$ from the main output port, as defined by (2.121a).

The combiner will combine the signals $y'_A(t)$ and $y_M(t)$ to produce the total output signal $y_T(t)$ given by

$$y_T(t) = \frac{y'_A(t) + y_M(t)}{\sqrt{2}} = \frac{1}{\sqrt{2}}\left[\frac{mW_A y_A(t)}{\sqrt{2}} + y_M(t)\right]$$ (2.123)

The total output signal $y_T(t)$ is now divided into two equal parts by a two-way divider.

One part is applied to the second input port of the multiplier through a gain block with gain G while the other part appears at the output port of the signal processing network as the final output signal $y_0(t)$ and is delivered to the input port of the radio receiver where $y_0(t)$ is given by

$$y_0(t) = \frac{y_T(t)}{\sqrt{2}} = \frac{1}{\sqrt{2}}\left[\frac{mW_A y_A(t)}{\sqrt{2}} + y_M(t)\right]$$ (2.124)

Steady-State Value of the Self-Adjusting Weighters

The multiplier has a conversion factor p with a dimension of volt^{-1}. It receives signals $Ay_A(t)/\sqrt{2}$ and $Gy_0(t)$ at its two input ports where $y_0(t)$ is converted into its complex conjugate. The product signal at the output of the multiplier is passed through a gain block with gain $-g$, an RC integrator circuit, a unity gain block, and finally is applied at the control port of the weighter W_A.

At the output of the gain block $-g$, the signal from the multiplier, is therefore given by

$$y_{om}(t) = -g \frac{[Apy_A(t)]}{\sqrt{2}} [Gy_o^*(t)] = -\frac{gpAG[y_A(t)y_o^*(t)]}{\sqrt{2}}$$

$$= -\mu W_A |y_A(t)|^2 - \frac{\sqrt{2}\mu[y_A(t)y_M^*(t)]}{m} \qquad (2.125)$$

$$\text{where} \qquad \mu = \frac{GAgpm}{4} \qquad (2.126)$$

and is the total gain factor associated with the feedback circuit controlling the weighter and has the dimension of power^{-1}. Also, p is the conversion factor of the multiplier and m is that of the weighter where both have the dimension of volt^{-1}.

Again, from (2.45), the signal at the input port of the RC integrator circuit is given by

$$y_{ii}(t) = \frac{TdW_A}{dt} + W_A \qquad (2.127)$$

Equating the two equations (2.125) and (2.127) gives

$$\frac{dW_A}{dt} = -\frac{[1 + \mu|y_A(t)|^2]W_A}{T} - \frac{\sqrt{2}\mu[y_A(t)y_M^*(t)]}{mT} \qquad (2.128)$$

At steady state, $dW_A/dt = 0$. Therefore,

$$[1 + \mu|y_A(t)|^2]W_{ASS} + \frac{\sqrt{2}\mu[y_A(t)y_M^*(t)]}{m} = 0 \qquad (2.129)$$

$$\text{or} \qquad W_{ASS} = -\frac{\sqrt{2}\mu[y_A(t)y_M^*(t)]}{m[1 + \mu|y_A(t)|^2]} \qquad (2.130)$$

Substitution of (2.121) into (2.130) gives the steady-state value of the weighter:

$$W_{ASS} = -\frac{\sqrt{2}[|G_A(\theta_1)G_M(\theta_1)|p_1 + |G_A(\theta_2)G_M(\theta_2)|p_2 + \ldots]}{m[(1/\mu) + |G_A(\theta_1)|^2 p_1 + |G_A(\theta_2)|^2 p_2 + \ldots]} \quad (2.131)$$

$$= -\frac{\sqrt{2}[|G_A(\theta_1)G_M(\theta_1)|p_1 + |G_A(\theta_2)G_M(\theta_2)|p_2 + \ldots]}{m[(1/\mu) + p_{A1} + p_{A2} + \ldots]}$$

where

$p_1 = |f_1(t)|^2$ = power of the interference signal 1;

$p_2 = |f_2(t)|^2$ = power of the interference signal 2;

$\cdots \qquad \cdots \qquad \cdots$

$p_{A1} = |G_A(\theta_1)|^2 p_1$ = power of interference signal 1 at AUX-A output port;

$p_{A2} = |G_A(\theta_2)|^2 p_2$ = power of interference signal 2 at AUX-A output port;

$\cdots \qquad \cdots \qquad \cdots$

2.7.3 Final Output Power

From (2.124), the final output power appearing at the output port of the signal processing network will be given by

$$p_o = |y_o(t)|^2 = \frac{1}{4}\left[\frac{mW_A y_A(t)}{\sqrt{2}} + y_M(t)\right]^2$$

Hence, the steady state output power p_{oSS} will be given by

$$p_{oSS} = \frac{1}{4}\left[\frac{m^2 W_{ASS}^2 |y_A(t)|^2}{2} + |y_M(t)|^2 + \frac{2m}{\sqrt{2}}\text{Re}\, W_{ASS} y_A(t) y_M(t)\right]$$
$$(2.132)$$

Substituting of (2.121) and (2.131) into (2.132) and simplifying, (Appendix E) gives

$$4p_{oSS} = p_{Md} + p_{M1}\left[1 - \frac{p_{A1} + K_{12}p_{A2} + K_{13}p_{A3} + \ldots}{(1/\mu) + p_{A1} + p_{A2} + p_{A3} + \ldots}\right]^2$$

$$+ p_{M2}\left[1 - \frac{K_{21}p_{A1} + p_{A2} + K_{23}p_{A3} + \ldots}{(1/\mu) + p_{A1} + p_{A2} + p_{A3} + \ldots}\right]^2 \qquad (2.133)$$

$$+ \ldots$$

$$= p_D + p_{\text{int1}} + p_{\text{int2}} + \ldots$$

where

$p_{Md} = |G_M(0)|^2 p_d$ = power of the desired signal at the main port;

$p_{M1} = |G_M(\theta_1)|^2 p_1$ = power of the interference signal 1 at the main port;

$p_{M2} = |G_M(\theta_2)|^2 p_2$ = power of the interference signal 2 at the main port;

$\cdots \qquad \cdots \qquad \cdots$

$p_d = |f_d(t)|^2$ = desired signal power.

Also,

$$K_{12} = \frac{1}{K_{21}} = \frac{|G_A(\theta_1)/G_A(\theta_2)|}{|G_M(\theta_1)/G_M(\theta_2)|}$$

$$K_{13} = \frac{1}{K_{31}} = \frac{|G_A(\theta_1)/G_A(\theta_3)|}{|G_M(\theta_1)/G_M(\theta_3)|}$$

$$K_{23} = \frac{1}{K_{32}} = \frac{|G_A(\theta_2)/G_A(\theta_3)|}{|G_M(\theta_2)/G_M(\theta_3)|}$$

$$\cdots \qquad \cdots \qquad \cdots$$

and $p_D = p_{Md}$ = desired signal power at the signal processing network output port.

$$p_{\text{int1}} = p_{M1}\left[1 - \frac{p_{A1} + K_{12}p_{A2} + K_{13}p_{A3} + \ldots}{(1/\mu) + p_{A1} + p_{A2} + p_{A3} + \ldots}\right]^2$$

$$= p_{M1}\left[\frac{1 + \mu p_{A2}(1 - K_{12}) + \mu p_{A3}(1 - K_{13}) + \ldots}{1 + \mu(p_{A1} + p_{A2} + p_{A3} + \ldots)}\right]^2 \qquad (2.134a)$$

$$p_{int2} = p_{M2}\left[1 - \frac{K_{21}p_{A1} + p_{A2} + K_{23}p_{A3} + \dots}{(1/\mu) + p_{A1} + p_{A2} + p_{A3} + \dots}\right]^2$$

$$= p_{M2}\left[\frac{1 + \mu p_{A1}(1 - K_{21}) + \mu p_{A3}(1 - K_{23}) + \dots}{1 + \mu(p_{A1} + p_{A2} + p_{A3} + \dots)}\right]^2 \qquad (2.134b)$$

$\dots \qquad \dots \qquad \dots$

where p_{int1} and p_{int2} are the powers of interference signals 1 and 2, respectively, contained in the final output power at the output port of the signal processing network.

2.7.4 Null Depth

From (2.134), the null depths corresponding to different interference signals will be given by

$$\eta_{int1} = \frac{[1 + \mu(p_{A1} + p_{A2} + p_{A3} + \dots)]^2}{[1 + \mu p_{A2}(1 - K_{12}) + \mu p_{A3}(1 - K_{13}) + \dots]^2} \qquad (2.135a)$$

$$\eta_{int2} = \frac{[1 + \mu(p_{A1} + p_{A2} + p_{A3} + \dots)]^2}{[1 + \mu p_{A1}(1 - K_{21}) + \mu p_{A3}(1 - K_{23}) + \dots]^2} \qquad (2.135b)$$

$\dots \qquad \dots \qquad \dots$

If the product of μ and the sum of the associated powers are greater than unity, then (2.135) will become

$$\eta_{int1} = \frac{[p_{A1} + p_{A2} + p_{A3} + \dots]^2}{[p_{A2}(1 - K_{12}) + p_{A3}(1 - K_{13}) + \dots]^2} \qquad (2.136a)$$

$$\eta_{int2} = \frac{[p_{A1} + p_{A2} + p_{A3} + \dots]^2}{[p_{A1}(1 - K_{21}) + p_{A3}(1 - K_{23}) + \dots]^2} \qquad (2.136b)$$

$\dots \qquad \dots \qquad \dots$

2.8 Multisource, Multioutput System

Here, the system under consideration contains a large number of transmitting signal sources while the associated beam former has a large number of active output ports.

2.8.1 The Signals

The elements of the linear array will pick up signals from all transmitting signal sources and deliver them to all input ports of the associated beam former. The beam former will then transport all signals to all of its output ports. From there the signals will pass through the active output ports of the beam former to enter the signal processing network and will be processed for subsequent removal of the interference signals.

2.8.1.1 Signal Source and Location

The block diagram of a typical multisource, multioutput adaptive system is shown in Figure 2.14. This is similar to the system discussed in Section 2.7

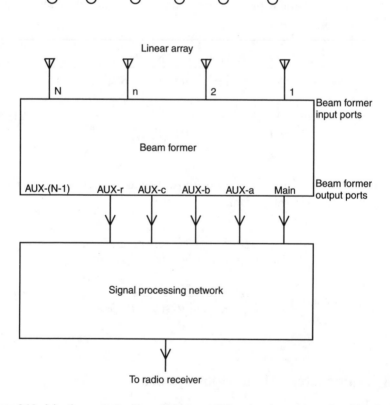

Figure 2.14 Adaptive system with multiple transmitting signal sources and multiple beam former output ports.

in the sense that this system also involves a large number of transmitting signal sources. Like before, a desired signal source is located along the array boresight of the main beam, transmitting the desired signal $f_d(t)$ and a number of interference signal sources denoted source 1, source 2, source 3, ..., located along θ_1, θ_2, θ_3, ..., off the array boresight and transmitting interference signals $f_1(t)$, $f_2(t)$, $f_3(t)$, ..., respectively. As before, it is assumed that none of the interference signal sources coincides with any of the AUX beams of the beam former. It is also assumed that all signals, desired or undesired, are uncorrelated.

However, unlike the system in Section 2.7, the beam former in this system contains a large number of active AUX output ports denoted by AUX-a, AUX-b, AUX-c, ...

2.8.1.2 Transfer Functions

The transfer function between the desired signal source and the main output port is denoted by $|G_M(0)|$.

Again, the transfer functions or gains between the interference signal sources and the main output port and those between the interference signal sources and the active AUX output ports are designated as follows:

For the main port: $\quad |G_M(\theta_1)|, |G_M(\theta_2)|, |G_M(\theta_3)|, \ldots$

For the AUX-a port: $\quad |G_a(\theta_1)|, |G_a(\theta_2)|, |G_a(\theta_3)|, \ldots$

For the AUX-b port: $\quad |G_b(\theta_1)|, |G_b(\theta_2)|, |G_b(\theta_3)|, \ldots$

For the AUX-c port: $\quad |G_c(\theta_1)|, |G_c(\theta_2)|, |G_c(\theta_3)|, \ldots$

$\cdots \qquad\qquad\qquad \cdots \qquad\quad \cdots$

For the AUX-r port: $\quad |G_r(\theta_1)|, |G_r(\theta_2)|, |G_r(\theta_3)|, \ldots$

$\cdots \qquad\qquad\qquad \cdots \qquad\quad \cdots$

2.8.1.3 Signals at the Beam Former Output Ports

The main output port of the beam former will contain signals from the desired signal source and all the interference signal sources. On the other hand, each AUX output port will contain signals from all interference signal sources and none from the desired signal source. Therefore, the signal $y_M(t)$ at the main output port and $y_a(t), y_b(t), y_c(t), \ldots$, at different AUX output ports AUX-a, AUX-b, AUX-c, ..., are given by

$$y_M(t) = y_{M1}(t) + y_{M2}(t) + y_{M3}(t) + \ldots + y_{Md}(t)$$
$$= |G_M(\theta_1)|f_1(t) + |G_M(\theta_2)|f_2(t) + |G_M(\theta_3)|f_3(t) + \ldots$$
$$+ |G_M(0)|f_d(t)$$
$$y_a(t) = y_{a1}(t) + y_{a2}(t) + y_{a3}(t) + \ldots$$
$$= |G_a(\theta_1)|f_1(t) + |G_a(\theta_2)|f_2(t) + |G_a(\theta_3)|f_3(t) + \ldots \quad (2.137)$$
$$y_b(t) = y_{b1}(t) + y_{b2}(t) + y_{b3}(t) + \ldots$$
$$= |G_b(\theta_1)|f_1(t) + |G_b(\theta_2)|f_2(t) + |G_b(\theta_3)|f_3(t) + \ldots$$
$$y_c(t) = y_{c1}(t) + y_{c2}(t) + y_{c3}(t) + \ldots$$
$$= |G_c(\theta_1)|f_1(t) + |G_c(\theta_2)|f_2(t) + |G_c(\theta_3)|f_3(t) + \ldots$$

$$\ldots \qquad \ldots \qquad \ldots$$

Each of the signals in (2.137) will be delivered to a corresponding input port of the signal processing network.

2.8.2 Signal Processing Network

Since the beam former has a large number of active output ports, the signal processing network will require an equally large number of input ports to receive those signals. One of these input ports is the main port and the rest are AUX ports. Thus, the signal processing network will process a large number of AUX port signals in order to remove a large number of interference signals.

2.8.2.1 Choice of Weighters

Since the signal processing network contains a large number of feedback loops, there will be an equally large number of associated weighters. Therefore, for such a system, manually adjustable weighters could not possibly be employed. Thus, all the weighters are likely to be self-adjusting.

2.8.2.2 Signal Processing Network with Self-Adjusting Weighters

As discussed above, all weighters used in the signal processing network will be automatic or self-adjusting. Thus, a block diagram of the signal processing network could be represented by Figure 2.10 where, for clarity, only the rth AUX port has been shown.

Total Output Signal

The signal from each AUX output port will be divided into two equal parts by a two-way divider. One part will be delivered to one of the two input ports of a multiplier through a gain block with gain A. The other part will be applied at the input port of a weighter. For easy identification, each multiplier and weighter pair is designated by its corresponding AUX output port. Thus, the rth feedback loop connected to the AUX-r output port will contain the rth multiplier and the rth weighter W_r.

At the output of the weighters, the signals will therefore be given by

$$y_a'(t) = \frac{mW_a y_a(t)}{\sqrt{2}} = \frac{mW_a}{\sqrt{2}} [y_{a1}(t) + y_{a2}(t) + y_{a3}(t) + \ldots]$$

$$y_b'(t) = \frac{mW_b y_b(t)}{\sqrt{2}} = \frac{mW_b}{\sqrt{2}} [y_{b1}(t) + y_{b2}(t) + y_{b3}(t) + \ldots] \quad (2.138)$$

$$y_c'(t) = \frac{mW_c y_c(t)}{\sqrt{2}} = \frac{mW_c}{\sqrt{2}} [y_{c1}(t) + y_{c2}(t) + y_{c3}(t) + \ldots]$$

$$\ldots \qquad \ldots \qquad \ldots$$

The signals, from the output ports of all the weighters, are then combined with the main port output signal $y_M(t)$ by an N-way combiner to give the total output signal $y_T(t)$ as

$$y_T(t) = \frac{y_a'(t) + y_b'(t) + y_c'(t) + \ldots + y_M(t)}{\sqrt{N}} \quad (2.139)$$

$$= \frac{1}{\sqrt{N}} \left[\frac{mW_a y_a(t)}{\sqrt{2}} + \frac{mW_b y_a(t)}{\sqrt{2}} + \frac{mW_c y_a(t)}{\sqrt{2}} + \ldots + y_M(t) \right]$$

The total output signal $y_T(t)$ is now divided into two equal parts by a two-way divider to give

$$y_o(t) = \frac{y_T(t)}{\sqrt{2}} \quad (2.140)$$

$$= \frac{1}{\sqrt{2N}} \left[\frac{mW_a y_a(t)}{\sqrt{2}} + \frac{mW_b y_a(t)}{\sqrt{2}} + \frac{mW_c y_a(t)}{\sqrt{2}} + \ldots + y_M(t) \right]$$

where one part will be delivered to the input port of an N-way divider and the other part will appear at the output port of the signal processing network as the final output signal to be delivered to the radio receiver.

Steady-State Values of the Self-Adjusting Weighters

Equation (2.140) is similar to (2.60), which is also the expression for the final output signal for a multiple output system. Referring to Figure 2.10, consider the AUX-r output port connected to the rth feedback loop containing the rth multiplier and the rth weighter W_r.

The rth multiplier will receive signals $[Ay_r(t)/\sqrt{2}]$ and $[Gy_o(t)/\sqrt{N}]$ at its two input ports. It will convert $y_o(t)$ to its complex conjugate and perform the multiplication process. As before, the product signal at the output of the rth multiplier will pass through a gain block $-g$, an RC integrating circuit, and a unity gain block before reaching the control port of the rth weighter (Figure 2.10). Therefore, the signals $y_{om}(t)$ at the output of the gain block $-g$ and $y_{ii}(t)$ at the input of the RC integrator will be given by (2.61 and 2.64)

$$y_{om}(t) = \frac{-gAGp}{\sqrt{(2N)}} [y_r(t)y_o^*(t)] \qquad (2.141a)$$

$$y_{ii}(t) = T(dW_r/dt) + W_r \qquad (2.141b)$$

Therefore, the time derivative of the rth weighter will be given by (2.65a)

$$\frac{dW_r}{dt} = -\frac{[1 + \mu|y_r(t)|^2]W_r}{T} - \frac{\mu y_r(t)}{T}[W_a y_a^*(t) + W_b y_b^*(t) + \ldots]$$

$$-\frac{\sqrt{2}\mu[y_r(t)y_M^*(t)]}{mT} \qquad (2.142)$$

where $\quad \mu = \dfrac{gAGpm}{(2\sqrt{2})N}$

is the total gain factor associated with the feedback circuit controlling the weighter and has the dimension of power^{-1}. As before, p and m are the conversion factors of the multiplier and the weighter, respectively, and have the dimension of volt^{-1}.

At steady state, $(dW_r/dt) = 0$; hence, substitution of (2.137) into (2.142) gives

$$[1 + \mu\{y_{r1}(t) + y_{r2}(t) + y_{r3}(t) + \ldots\}^2]W_{rSS}$$
$$+ \mu[y_{r1}(t) + y_{r2}(t) + \ldots][W_{aSS}\{y_{a1}(t) + y_{a2}(t) + \ldots\} \quad (2.143)$$
$$+ W_{bSS}\{y_{a1}(t) + y_{a2}(t) + \ldots\}]$$
$$+ \frac{\sqrt{2}\mu}{m}[y_{r1}(t) + y_{r2}(t) + \ldots][y_{r1}(t) + y_{r2}(t) + \ldots] = 0$$

Therefore, to obtain the steady-state value of each weighter, solution of a large number of simultaneous equations similar to (2.143) is necessary where the number of equations to be solved is equal to the number of active AUX output ports and hence the number of feedback loops.

Obviously, the mathematics involved will be quite complicated and hence an alternate means to estimate the null depth corresponding to each interference signal will have to be determined.

2.8.3 General Expression for Null Depth

A close examination of (2.118) and (2.135) reveals that a general expression for null depth involving multiple interference signals and multiple active AUX output ports could be written as

$$\eta_1 = \frac{[1 + \mu\{p_{a1} + p_{b1} + p_{c1} + \ldots p_{a2} + p_{b2} + p_{c2} + \ldots \\ p_{a3} + p_{b3} + p_{c3} + \ldots\}]^2}{[1 + \mu\{p_{a1}(1 - K_{a11}) + p_{a2}(1 - K_{a12}) + p_{a3}(1 - K_{a13}) + \ldots \\ p_{b1}(1 - K_{b11}) + p_{b2}(1 - K_{b12}) + p_{b3}(1 - K_{b13}) + \ldots \\ p_{c1}(1 - K_{c11}) + p_{c2}(1 - K_{c12}) + p_{c3}(1 - K_{c13}) + \ldots\}]^2}$$

$$(2.144a)$$

$$\eta_2 = \frac{[1 + \mu\{p_{a1} + p_{b1} + p_{c1} + \ldots p_{a2} + p_{b2} + p_{c2} + \ldots \\ p_{c1} + p_{c2} + p_{c3} + \ldots\}]^2}{[1 + \mu\{p_{a1}(1 - K_{a21}) + p_{a2}(1 - K_{a22}) + p_{a3}(1 - K_{a23}) + \ldots \\ p_{b1}(1 - K_{b21}) + p_{b2}(1 - K_{b22}) + p_{b3}(1 - K_{b23}) + \ldots \\ p_{c1}(1 - K_{c21}) + p_{c2}(1 - K_{c22}) + p_{c3}(1 - K_{c23}) + \ldots\}]^2}$$

$$(2.144b)$$

$$\eta_3 = \frac{[1 + \mu\{p_{a1} + p_{b1} + p_{c1} + \dots p_{a2} + p_{b2} + p_{c2} + \dots \\ p_{c1} + p_{c2} + p_{c3} + \dots \}]^2}{[1 + \mu\{p_{a1}(1 - K_{a31}) + p_{a2}(1 - K_{a32}) + p_{a3}(1 - K_{a33}) + \dots \\ p_{b1}(1 - K_{b31}) + p_{b2}(1 - K_{b32}) + p_{b3}(1 - K_{b33}) + \dots \\ p_{c1}(1 - K_{c31}) + p_{c2}(1 - K_{c32}) + p_{c3}(1 - K_{c33}) + \dots \}]^2}$$

$$\dots \qquad \dots \qquad \dots$$

(2.144c)

where

$p_{a1} = |G_a(\theta_1)|^2 |f_1(t)|^2$ = power at AUX-a port due to interference signal 1;

$p_{b1} = |G_b(\theta_1)|^2 |f_1(t)|^2$ = power at AUX-b port due to interference signal 1;

$p_{c1} = |G_c(\theta_1)|^2 |f_1(t)|^2$ = power at AUX-c port due to interference signal 1;

$$\dots \qquad \dots \qquad \dots$$

$p_{a2} = |G_a(\theta_1)|^2 |f_1(t)|^2$ = power at AUX-a port due to interference signal 2;

$p_{b2} = |G_b(\theta_1)|^2 |f_1(t)|^2$ = power at AUX-b port due to interference signal 2;

$p_{c2} = |G_c(\theta_1)|^2 |f_1(t)|^2$ = power at AUX-c port due to interference signal 2;

$$\dots \qquad \dots \qquad \dots$$

$p_{a3} = |G_a(\theta_1)|^2 |f_1(t)|^2$ = power at AUX-a port due to interference signal 3;

$p_{b3} = |G_b(\theta_1)|^2 |f_1(t)|^2$ = power at AUX-b port due to interference signal 3;

$p_{c3} = |G_c(\theta_1)|^2 |f_1(t)|^2$ = power at AUX-c port due to interference signal 3;

$$\dots \qquad \dots \qquad \dots$$

and

$$K_{rmn} = \frac{1}{K_{rnm}} = \frac{|G_r(\theta_m)/G_r(\theta_n)|}{|G_M(\theta_m)/G_M(\theta_n)|}$$

(2.145)

and where $r = a, b, c, \ldots$; $m = 1, 2, 3, \ldots$; and $n = 1, 2, 3, \ldots$. Note that

$$K_{rmm} = K_{rnn} = 1 \tag{2.146}$$

Thus, using (2.144), the null depth corresponding to a system with any number of active output ports, involving any number of interference signal sources could readily be estimated.

Example 2.12

Using (2.144), derive expressions for null depths for systems with (a) one interference signal source where the system beam former has one active AUX output port; (b) one interference signal source where the system beam former has two active AUX output ports; (c) one interference signal source where the system beam former has multiple active AUX output ports; (d) two interference signal sources where the system beam former has one active AUX output port; (e) two interference signal sources where the system beam former has two active AUX output ports; and (f) multiple interference signal sources where the system beam former has one active AUX output port.

Answer

(a) There is one interference signal source 1 and one active AUX output port, AUX-a. Since only one interference signal is to be nulled, only η_1 is involved. Therefore, only (2.144a) will have to be considered where the quantities not containing the combination $(a1)$ will vanish. Hence, from (2.144a),

$$\eta_1 = \frac{[1 + \mu p_{a1}]^2}{[1 + \mu p_{a1}(1 - K_{a11})]^2} = [1 + \mu p_{a1}]^2$$

This is the same as (2.34).

(b) There is one interference signal source, 1, and two active AUX output ports, AUX-a and AUX-b. As before, since there is only one interference signal source, only η_1 will be involved. However, in this case, quantities not having $(a1)$ and $(b1)$ will vanish. So, from (2.144a),

$$\eta_1 = \frac{[1 + \mu(p_{a1} + p_{b1})]^2}{[1 + \mu p_{a1}(1 - K_{a11}) + \mu p_{b1}(1 - K_{b11})]^2} = [1 + \mu(p_{a1} + p_{b1})]^2$$

This is the same as (2.55).

(c) In this case there is still only one interference signal source 1, but the beam former has multiple AUX output ports, for example, AUX-a, AUX-b, AUX-c, and so on. Hence, in (2.144a), all quantities not having $(a1)$, $(b1)$, $(c1)$, . . . , will vanish. Thus, from (2.144a),

$$\eta_1 = \frac{[1 + \mu(p_{a1} + p_{b1} + p_{c1} \ldots)]^2}{[1 + \mu\{p_{a1}(1 - K_{a11}) + p_{b1}(1 - K_{b11}) + p_{c1}(1 - K_{c11}) + \ldots\}]^2}$$

$$= [1 + \mu(p_{a1} + p_{b1} + p_{c1})]^2$$

Again, this is the same as (2.78).

(d) In this case there are two interference signal sources, 1 and 2, and the beam former has only one AUX output port, AUX-a. Thus, there are two interference signals to be nulled, η_1 and η_2. Hence, in (2.144a) and (2.144b), all quantities not having $(a1)$ and $(a2)$ will vanish. Hence,

$$\eta_1 = \frac{[1 + \mu(p_{a1} + p_{a2})]^2}{[1 + \mu\{p_{a1}(1 - K_{a11}) + p_{a2}(1 - K_{a12})\}]^2} = \frac{[1 + \mu(p_{a1} + p_{a2})]^2}{[1 + \mu p_{a2}(1 - K_{a12})]^2}$$

$$\eta_2 = \frac{[1 + \mu(p_{a1} + p_{a2})]^2}{[1 + \mu\{p_{a1}(1 - K_{a21}) + p_{a2}(1 - K_{a22})\}]^2} = \frac{[1 + \mu(p_{a1} + p_{a2})]^2}{[1 + \mu p_{a1}(1 - K_{a21})]^2}$$

These two expressions are the same as (2.98a) and (2.98b).

(e) Here, as in (d), there are two interference signal sources, source 1 and source 2. However, in this case, the beam former has two active output ports, AUX-a and AUX-b. Therefore, in (2.144a) and (2.144b), all the quantities not having $(a1)$, $(a2)$, $(b1)$, and $(b2)$ will vanish. Thus,

$$\eta_1 = \frac{[1 + \mu(p_{a1} + p_{a2} + p_{b1} + p_{b2})]^2}{[1 + \mu\{p_{a1}(1 - K_{a11}) + p_{a2}(1 - K_{a12}) + p_{b1}(1 - K_{b11}) + p_{b2}(1 - K_{b12})\}]^2}$$

$$= \frac{[1 + \mu(p_{a1} + p_{a2} + p_{b1} + p_{b2})]^2}{[1 + \mu\{p_{a2}(1 - K_{a12}) + p_{b2}(1 - K_{b12})\}]^2}$$

$$\eta_2 = \frac{[1 + \mu(p_{a1} + p_{a2} + p_{b1} + p_{b2})]^2}{[1 + \mu\{p_{a1}(1 - K_{a21}) + p_{a2}(1 - K_{a22}) + p_{b1}(1 - K_{b21}) + p_{b2}(1 - K_{b22})\}]^2}$$

$$= \frac{[1 + \mu(p_{a1} + p_{a2} + p_{b1} + p_{b2})]^2}{[1 + \mu\{p_{a1}(1 - K_{a21}) + p_{b1}(1 - K_{b21})\}]^2}$$

the two expressions for η_1 and η_2 are the same as (2.108a) and (2.108b).

(f) In this case, the system contains a multiple number of interference signal sources, source 1, source 2, source 3, and so on, but the beam former has only one active AUX output port, AUX-a. Therefore, in (2.144), all the quantities not having (a1), (a2), (a3), . . . , will vanish. Hence, (2.144) gives

$$\eta_2 = \frac{[1 + \mu(p_{a1} + p_{a2} + p_{a3} + \dots)]^2}{[1 + \mu\{p_{a1}(1 - K_{a11}) + p_{a2}(1 - K_{a12}) + p_{a3}(1 - K_{a13}) + \dots\}]^2}$$

$$= \frac{[1 + \mu(p_{a1} + p_{a2} + p_{a3} + \dots)]^2}{[1 + \mu\{p_{a2}(1 - K_{a12}) + p_{a3}(1 - K_{a13}) + \dots\}]^2}$$

$$\eta_2 = \frac{[1 + \mu(p_{a1} + p_{a2} + p_{a3} + \dots)]^2}{[1 + \mu\{p_{a1}(1 - K_{a21}) + p_{a2}(1 - K_{a22}) + p_{a3}(1 - K_{a23}) + \dots\}]^2}$$

$$= \frac{[1 + \mu(p_{a1} + p_{a2} + p_{a3} + \dots)]^2}{[1 + \mu\{p_{a1}(1 - K_{a21}) + p_{a3}(1 - K_{a23}) + \dots\}]^2}$$

These expressions for null depths are the same as in (2.135).

2.8.4 Alternate Expressions for Null Depth

As discussed in Section 2.4, for a multisource multiport system, the sum of the AUX port output powers, for different AUX output ports, can be expressed as

$$\lambda_1 = p_{a1} + p_{b1} + p_{c1} + \dots$$
$$\lambda_2 = p_{a2} + p_{b2} + p_{c2} + \dots$$
$$\lambda_3 = p_{a3} + p_{b3} + p_{c3} + \dots$$

$$\dots \quad \dots \quad \dots$$

Then (2.144) can be written as

$$\eta_1 = \frac{[1 + \mu\{\lambda_1 + \lambda_2 + \lambda_3 + \dots \}]^2}{\begin{array}{l}[1 + \mu\{p_{a1}(1 - K_{a11}) + p_{a2}(1 - K_{a12}) + p_{a3}(1 - K_{a13}) + \dots \\ p_{b1}(1 - K_{b11}) + p_{b2}(1 - K_{b12}) + p_{b3}(1 - K_{b13}) + \dots \\ p_{c1}(1 - K_{c11}) + p_{c2}(1 - K_{c12}) + p_{c3}(1 - K_{c13}) + \dots \}]^2 \end{array}}$$

(2.147a)

$$\eta_2 = \frac{[1 + \mu\{\lambda_1 + \lambda_2 + \lambda_3 + \dots \}]^2}{\begin{array}{l}[1 + \mu\{p_{a1}(1 - K_{a21}) + p_{a2}(1 - K_{a22}) + p_{a3}(1 - K_{a23}) + \dots \\ p_{b1}(1 - K_{b21}) + p_{b2}(1 - K_{b22}) + p_{b3}(1 - K_{b23}) + \dots \\ p_{c1}(1 - K_{c21}) + p_{c2}(1 - K_{c22}) + p_{c3}(1 - K_{c23}) + \dots \}]^2 \end{array}}$$

(2.147b)

$$\eta_3 = \frac{[1 + \mu\{\lambda_1 + \lambda_2 + \lambda_3 + \dots \}]^2}{\begin{array}{l}[1 + \mu\{p_{a1}(1 - K_{a31}) + p_{a2}(1 - K_{a32}) + p_{a3}(1 - K_{a33}) + \dots \\ p_{b1}(1 - K_{b31}) + p_{b2}(1 - K_{b32}) + p_{b3}(1 - K_{b33}) + \dots \\ p_{c1}(1 - K_{c31}) + p_{c2}(1 - K_{c32}) + p_{c3}(1 - K_{c33}) + \dots \}]^2 \end{array}}$$

(2.147c)

2.8.5 The Tradeoff

The null depth corresponding to a given undesired signal depends, to a large extent, on the location of the source of the undesired signal relative to the locations of the AUX beams of the beam former. This would provide the means for a possible tradeoff.

If a certain reduction in the level of the desired signal power can be tolerated, then the orientation of the array can be adjusted to increase the null depth. This would reduce the power level of the desired signal, but the improvement in the null depth may be substantial. A compromise could therefore be made between the reduction of the desired power level and the increase in null depth.

2.8.6 Number of Undesired Signals and the Active Output Ports

It has been shown in Section 2.3 that if the number of active output ports is larger than the number of undesired signals involved, then the contributions to the null depths by the excess active output ports are quite small.

On the other hand, Section 2.5 shows that if the number of undesired signal sources is larger than the number of active output ports, considerable nulling for all the undesired signals is still possible, provided the resulting null depths are acceptable.

Thus, for a given system with a certain number of active output ports, computer software could be developed that would quickly calculate the achievable null depth for each undesired signal using the location of its source relative to the AUX beams of the beam former. In many applications, the resulting null depths would probably be sufficient. If necessary, the orientation of the array could be adjusted as mentioned above.

2.9 Leakage of the Desired Signal into the AUX Ports

With an ideal beam former, when a signal source is placed along the array boresight or the main beam, the resulting signal will appear only at the main output port and not at any of the AUX output ports. However, in the real world, there is no such thing as an ideal beam former. Therefore, in a practical beam former, a fraction of the desired signal will leak into other ports and, ultimately, will appear at every AUX output port.

2.9.1 The Leakage

In a system where there is a desired signal source located along the array boresight and an arbitrary number of interference signal sources distributed randomly on both sides of the boresight, the signal at the output of a given AUX port will ideally contain only the interference signals and no desired signal. However, in the presence of leakage, the AUX port signals will also contain a small fraction of the desired signal.

In this section, the effect of the leakage of the desired signal into an AUX output port will be demonstrated with the help of a simple system having two transmitting signal sources and two active beam former output ports, as represented by Figure 2.4.

2.9.2 The Signals

Like the system discussed in Section 2.2, the present system contains two transmitting signal sources. One of them is the desired signal source located along the array boresight and transmitting the desired signal $f_d(t)$. The other is an interference signal source located at an arbitrary direction θ_i off the boresight and transmitting the interference signal $f_i(t)$. It is assumed that the direction θ_i does not coincide with any of the AUX beams of the beam former. It is also assumed that $f_d(t)$ and $f_i(t)$ are uncorrelated.

Also, the beam former has only two active output ports. One is the main output port while the other is an arbitrary AUX output port denoted the AUX-A port.

As before, the transfer function or gain between the main output port and the desired signal source will be designated $|G_M(0)|$ while that between the main output port and the interference signal source will be designated $|G_M(\theta_i)|$.

Similarly, the transfer function or gain between the AUX-A output port and the interference signal source will be designated $|G_A(\theta_i)|$.

However, in the situation under discussion, in addition to the interference signal, the AUX port will also contain part of the desired signal. This will introduce an extra transfer function or gain that exists between the AUX-A output port and the desired signal source located along the boresight ($\theta = 0$) and hence this gain function will be designated $|G_A(0)|$.

Therefore, at the output of the beam former, the main port signal $y_M(t)$ and the AUX-A port signal $y_A(t)$ are given by

$$y_M(t) = |G_M(0)|f_d(t) + |G_M(\theta_i)|f_i(t) \qquad (2.148\text{a})$$

$$y_A(t) = |G_A(0)|f_d(t) + |G_A(\theta_i)|f_i(t) \qquad (2.148\text{b})$$

Signals $y_M(t)$ and $y_A(t)$ are now delivered to the corresponding input ports of the signal processing network.

2.9.3   Signal Processing Network

As mentioned above, for the purpose of demonstration, a system with two transmitting signal sources and an associated beam former with two active output ports is being considered here. Thus, the signal processing network will have only two input ports, the main input port and an AUX output port to receive the signals from the beam former.

The AUX input port is connected to a feedback loop containing an associated weighter, which is assumed to be automatic or self-adjusting.

2.9.3.1   Total Output Signal

Then, the signal $y_A(t)$ at the AUX-A output port will be divided into two equal parts where one part ($y_A(t)/\sqrt{2}$) will be delivered to a gain block with gain A and the other part will be delivered to the weighter W_A. The output $[Ay_A(t)/\sqrt{2}]$ of gain block A will, in turn, be delivered to one of the two input ports of a multiplier and the output $y'_A(t)$ of the weighter W_A will be delivered to one of the two input ports of a two-way combiner. The signal $y'_A(t)$, appearing at the output of the weighter W_A, is given by

$$y'_A(t) = \frac{mW_A y_A(t)}{\sqrt{2}} = \frac{mW_A[|G_A(0)|f_d(t) + |G_A(\theta_i)|f_i(t)]}{\sqrt{2}}$$

(2.149)

where m is the conversion factor of the weighter having the dimension of volt^{-1}. The main port signal $y_M(t)$ is applied at the other input port of the two-way combiner, which combines $y'_A(t)$ with $y_M(t)$ to give the total output signal $y_T(t)$ as

$$y_T(t) = \frac{y'_A(t) + y_M(t)}{\sqrt{2}} = \frac{1}{\sqrt{2}}\left[\frac{mW_A\{|G_A(0)|f_d(t) + |G_A(\theta_i)|f_i(t)\}}{\sqrt{2}} \right.$$
$$\left. + \{|G_M(0)|f_d(t) + |G_M(\theta_i)|f_i(t)\} \right]$$

(2.150)

Signal $y_T(t)$ will now be divided into two equal parts to give the final output signal $y_o(t)$ as

$$y_o(t) = \frac{y_T(t)}{\sqrt{2}} = \frac{y'_A(t) + y_M(t)}{2} = \frac{1}{2}\left[\frac{mW_A\{|G_A(0)|f_d(t) + |G_A(\theta_i)|f_i(t)\}}{\sqrt{2}} \right.$$
$$\left. + \{|G_M(0)|f_d(t) + |G_M(\theta_i)|f_i(t)\} \right]$$

(2.151)

where one part will be delivered to the second input port of the multiplier through a gain block with gain G and the other part will appear at the output port of the signal processing network and will subsequently be delivered to the input port of a radio receiver.

2.9.3.2 Steady-State Value of the Weighter

The multiplier has a conversion factor p with the dimension of volt^{-1}. As mentioned above, it receives signals $[Ay_A(t)/\sqrt{2}]$ and $[Gy_o(t)]$ at its two input ports. It converts $y_o(t)$ into its complex conjugate $y_o^*(t)$ and performs the multiplication.

The product signal at the output of the multiplier is passed through a gain block with gain $-g$, an RC integrating circuit, a unity gain block, and finally is applied at the control input port of the weighter.

As discussed in Section 2.2, the signal $y_{om}(t)$ at the output of the multiplier after the gain block $-g$ is (2.23)

$$y_{om}(t) = -\mu W_A |y_A(t)|^2 + \frac{\sqrt{2}\mu y_A(t) y_M^*(t)}{m} \qquad (2.152)$$

where

$$\mu = \frac{Gagpm}{4}$$

and represents the total gain factor associated with the feedback circuit controlling the weighter. It has the dimension of power^{-1}. Again, the signal $y_{ii}(t)$ at the input port of the RC integrator is (2.25)

$$y_{ii}(t) = T\frac{dW_A}{dt} + W_A \qquad (2.153)$$

where $T = RC$. Since $y_{om}(t)$ and $y_{ii}(t)$ represent the same signal, (2.152) and (2.153) give

$$\frac{dW_A}{dt} = -\frac{[1 + \mu|y_A(t)|^2]W_A}{T} - \frac{\sqrt{2}\mu y_A(t) y_M^*(t)}{mT} \qquad (2.154)$$

which at steady state gives (2.26)

$$W_{ASS} = -\frac{\sqrt{2}\mu}{m} \frac{|y_A(t) y_M^*(t)|}{1 + \mu|y_A(t)|^2} \qquad (2.155)$$

Keeping in mind that signals $f_d(t)$ and $f_i(t)$ are uncorrelated, then substituting (2.148) into (2.155) and simplifying give

$$W_{ASS} = -\frac{\sqrt{2}}{m} \frac{\mu|G_A(0)G_M(0)|p_d + \mu|G_A(\theta_i)G_M(\theta_i)|p_i}{1 + \mu|G_A(0)|^2 p_d + \mu|G_A(\theta_i)|^2 p_i} \qquad (2.156)$$

where $p_d = |f_d(t)|^2$ = power of the desired signal and $p_i = |f_i(t)|^2$ = power of the interference signal.

2.9.4 Final Output Power

The final output power, measured at the output port of the signal processing network is obtained from (2.151) as

$$p_o = |y_o(t)|^2 = \frac{1}{4}\left[\frac{mW_A[|G_A(0)|f_d(t) + |G_A(\theta_i)|f_i(t)]}{2}\right.$$

$$\left. + [|G_M(0)|f_d(t) + |G_M(\theta_i)|f_i(t)]\right]^2 \tag{2.157a}$$

or

$$4p_o = \frac{m^2 W_A^2[|G_A(0)|f_d(t) + |G_A(\theta_i)|f_i(t)]^2}{2}$$

$$+ [|G_M(0)|f_d(t) + |G_M(\theta_i)|f_i(t)]^2 \tag{2.157b}$$

$$+ \frac{2m}{\sqrt{2}}W_A[|G_A(0)|f_d(t) + |G_A(\theta_i)|f_i(t)][y_M(t)]$$

At steady state, $W_A = W_{ASS}$. Therefore, substituting (2.156) into (2.157b) and simplifying, the steady-state final output power p_{oSS} is given by (Appendix F)

$$4p_{oSS} = p_{Md}\frac{[1 + \mu p_{Ai}(1 - K_{A01})]^2}{[1 + \mu p_{Ai} + \mu p_{Ad}]^2} \tag{2.158a}$$

$$+ p_{Mi}\frac{[1 + \mu p_{Ad}(1 - K_{A10})]^2}{[1 + \mu p_{Ai} + \mu p_{Ad}]^2} \tag{2.158b}$$

$$= p_D + p_{int}$$

where p_D represents the desired signal power and p_{int} represents the interference signal power contained in the total output power.

From (2.158a), the desired signal power p_D is given by

$$p_D = p_{Md}\frac{[1 + \mu p_{Ai}(1 - K_{A01})]^2}{[1 + \mu p_{Ai} + \mu p_{Ad}]^2} \tag{2.159a}$$

$$= p_{Md}F_d \tag{2.159b}$$

where p_{Md} is the desired signal power appearing at the main output port and F_d is a modification factor associated with the desired signal, given by

$$F_d = \frac{[1 + \mu p_{Ai}(1 - K_{A01})]^2}{[1 + \mu p_{Ai} + \mu p_{Ad}]^2} \tag{2.160}$$

Again, from (2.158a), the interference signal power p_{int} is given by

$$p_{\text{int}} = p_{Mi} \frac{[1 + \mu p_{Ad}(1 - K_{A10})]^2}{[1 + \mu p_{Ai} + \mu p_{Ad}]^2} \tag{2.161}$$

$$\text{where} \quad K_{A01} = \frac{1}{K_{A10}} = \frac{|G_A(0)/|G_A(\theta_i)|}{|G_M^*(0)/G_M^*(\theta_i)|} \tag{2.162}$$

From (2.159) we can see that if there is no leakage of the desired signal into the AUX port, then $G_A(0)$ will be zero, and hence both p_{Ad} and K_{A01} will be zero. Then F_d will be unity and p_D will be equal to p_{Md} as shown in (2.33a).

2.9.5 Null Depth

From (2.161), the expected null depth η is given by

$$\eta = \frac{p_{Mi}}{p_{\text{int}}} = \frac{[1 + \mu p_{Ai} + \mu p_{Ad}]^2}{[1 + \mu p_{Ad}(1 - K_{A10})]^2}$$

$$= [1 + \mu p_{Ai}]^2 \frac{\left[1 + \dfrac{\mu p_{Ad}}{1 + \mu p_{Ai}}\right]^2}{[1 + \mu p_{Ad}(1 - K_{A10})]^2} \tag{2.163}$$

$$= \eta_0 F_n$$

$$\text{where} \quad \eta_0 = [1 + \mu p_{Ai}]^2 \tag{2.164a}$$

is the null depth without any leakage from the desired signal source as given by (2.34a) and F_n is the null modification factor given by

$$F_n = \frac{\left[1 + \dfrac{\mu p_{Ad}}{1 + \mu p_{Ai}}\right]^2}{[1 + \mu p_{Ad}(1 - K_{A10})]^2} \tag{2.164b}$$

Example 2.13

An adaptive array system utilizes a 4×4 beam former. There is a desired signal source located along the array boresight and an interference signal source, of equal strength, is located along $+20°$ off the boresight. The main and AUX-(+1) ports of the beam former are active while the other two AUX ports are match terminated. The corresponding null depth is 30 dB without any leakage of the desired signal into the AUX port. Now, let there be a leakage from the desired signal source into the AUX port such that the normalized gain function between the desired signal source and the AUX-(+1) output port is 0.05.

Determine the effect of this leakage on the final desired signal output power and the resulting null depth.

Answer

Given

$$p_i = p_d$$

$$\eta_0 = [1 + \mu p_{(+1)i}]^2 = 30 \text{ dB}$$

$$|G_{(+1)}(0°)| = 0.05$$

Again from Figure 1.6

$$|G_M(0)| = 1.0000$$

$$|G_M(20°)| = 0.4091$$

$$|G_{(+1)}(20°)| = 0.8524$$

Now,

$$\mu p_{(+1)i} = \mu |G_{(+1)}(20°)|^2 p_i = 30.6228$$

and

$$\mu p_{(+1)d} = \mu |G_{(+1)}(0°)|^2 p_d$$

Therefore,

$$\mu p_{(+1)d} = 0.1054$$

Also,

$$K_{(+1)01} = 1/K_{(+1)10} = 0.24$$

Therefore,

$$F_d = -2.33 \text{ dB}$$

and

$$F_n = +3.56 \text{ dB}$$

Thus, although the leakage improves the null depth by about 3.5 dB, the desired signal power is reduced by more than 2 dB.

Example 2.14

Calculate and plot the values of F_d and F_n for various amounts of leakage of the desired signal into the AUX port.

Answer

The different parameters required to calculate F_d and F_n are shown in Table 2.12 along with the corresponding values of F_d and F_n. A plot of F_d and F_n, as a function of $|G_{(+1)}(0)|$, is shown in Figure 2.15. Figure 2.15 shows that the desired signal power decreases with increasing leakage. However, as the leakage increases, the null depth first increases to a maximum value of about 5.5 dB and then decreases.

2.9.6 Minimization of Leakage of the Desired Signal into an AUX Port

From the preceding example, it is evident that a small amount of leakage of the desired signal into the AUX port can cause a considerable degradation of the output power level of the desired signal. Necessary precautions should therefore be taken to prevent such phenomena.

Table 2.12

$|G_{(+1)}(0)|$	0.00	0.02	0.04	0.06	0.08	0.10	0.12	0.14
$K_{(+1)01}$	0.000	0.096	0.192	0.288	0.384	0.480	0.576	0.672
$K_{(+1)10}$	0.000	10.418	5.208	3.473	2.605	2.084	1.736	1.488
$\mu p_{(+1)d}$	0.000	0.017	0.067	0.152	0.270	0.341	0.422	0.826
F_d (dB)	0.00	−0.85	−1.81	−2.88	−4.11	−5.55	−7.25	−9.36
F_n (dB)	0.00	+1.52	+2.90	+4.13	+5.00	+5.41	+5.31	+4.71

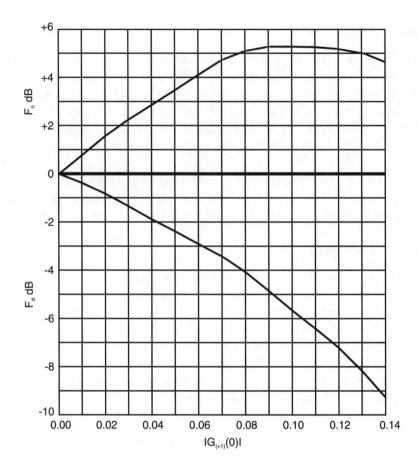

Figure 2.15 Variation of the desired signal modification factor F_d and the null depth modification factor F_n with the leakage $|G_{(+1)}(0)|$ of the desired signal into the AUX-(+1) port.

The beams are generated by the beam former and hence care must be taken during the design of the beam former to ensure that the signals traveling through these devices are properly balanced.

The design procedure and the characteristics of a beam former have been discussed in detail in Section 2.13.

2.10 Effect of Offset Voltage on System Performance

The offset voltage associated with a particular circuit device could affect the overall performance of a given system. This happens when the offset voltage combines with the signal voltage to produce erroneous results.

2.10.1   Offset Voltage

The offset voltages are usually associated with the multipliers and the weighters. However, if a multiplier is ac coupled to a weighter, then the offset voltage will be blocked and the source of any offset voltage will only be the weighter. The effect of any offset voltage associated with a weighter will be to modify the signal at the output port of the weighter. This will affect the final output power at the output port of the signal processing network.

The schematic diagram of a typical weighter is shown in Figure 2.16 along with its phasor diagram without and with offset voltage. The voltage gain and phase as a function of control voltage are presented in Table 2.13.

2.10.2   The Signals

For the purpose of demonstration, the basic two-source two-output adaptive system discussed in Section 2.2 will be considered here.

As discussed in Section 2.2, of the two transmitting signal sources, one is the desired signal source located along the array boresight and transmitting the desired signal $f_d(t)$, while the other is an interference signal source located in an arbitrary direction of θ_i off the array boresight and transmitting the interference signal $f_i(t)$. The direction θ_i does not coincide with any of the AUX beams produced by the beam former. Moreover, the signals $f_d(t)$ and $f_i(t)$ are uncorrelated.

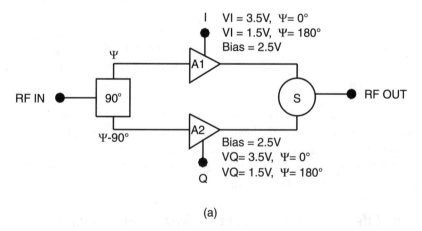

(a)

Figure 2.16 (a) Schematic diagram of a typical weighter, (b) complex weighter phasor diagram without offset voltage, and (c) complex weighter phasor diagram with offset voltage.

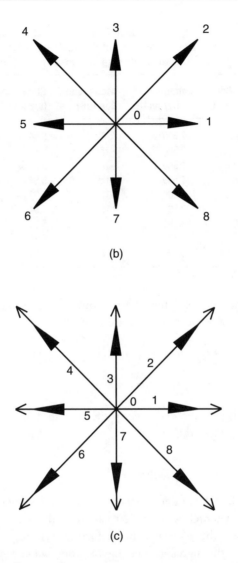

(b)

(c)

Figure 2.16 (continued).

Again, of the two active output ports of the beam former, one is the main port and the other is an arbitrary AUX port designated AUX-A. All other output ports are match terminated.

Now, let $|G_M(0)|$ be the transfer functions or gains between the main output port and the desired signal source, while $|G_M(\theta_i)|$ is that between the main output port and the interference signal source. Also, let $|G_A(\theta_i)|$ be the transfer function or gain between the AUX-A output port and the

Table 2.13
Voltage Gain (dB) and Phase as Functions of Control Voltage for a Typical Weighter
(No Offset Voltage Is Involved)

Vector Orientation	Control Voltage $v(I)$ (volt)	$v(Q)$ (volt)	Magnitude Log (dB)	Slope Lin (v/v × v)	Phase (Deg)
0	2.500	2.500	<−85	5.94	xxx
1	3.500	2.500	+15	5.94	0
2	3.500	1.500	+18	5.94	+45
3	2.500	1.500	+15	5.94	+90
4	3.500	1.500	+18	5.94	+135
5	3.500	2.500	+15	5.94	180
6	3.500	3.500	+18	5.94	−135
7	2.500	3.500	+15	5.94	−90
8	3.500	3.500	+18	5.94	−45

interference signal source. Then the two output signals $y_M(t)$ and $y_A(t)$ at the main and AUX-A output ports are given by

$$y_M(t) = |G_M(0)|f_d(t) + |G_M(\theta_i)|f_i(t) \qquad (2.165a)$$

$$y_A(t) = |G_A(\theta_i)|f_i(t) \qquad (2.165b)$$

These two signals are then delivered to the corresponding input ports of the signal processing network.

2.10.3   Signal Processing Network

Here again, for the sake of demonstration, the simplest model of an adaptive array system has been chosen where there are only two transmitting signal sources and where the associated beam former has only two active output ports. Therefore, the signal processing network will have only two input ports, namely, the main port and an AUX port, which will be connected to a feedback loop in association with a weighter. It is assumed that the weighter is an automatic or self-adjusting one.

2.10.3.1   Final Output Signal

Referring to Figure 2.4, signal $y_A(t)$ at the AUX-A output port will be divided into two equal parts $[y_A(t)/\sqrt{2}]$ by a two-way divider where one part will be delivered to the input port of a gain block with gain A and the other part will be delivered to the input port of the weighter W_A. The output $[Ay_A(t)/\sqrt{2}]$ of the gain block will be applied at one of the two input ports of a multiplier while the output $y_A'(t)$ of the weighter will be applied at one of the two input ports of a two-way combiner.

However, there is an offset voltage V_{AOFF} associated with weighter W_A. Hence, at the output port of the weighter, the signal $y_A'(t)$ will be given by

$$y_A'(t) = \frac{m[W_A + V_{AOFF}]y_A(t)}{\sqrt{2}} = \frac{mW_{AOFF}\,y_A(t)}{\sqrt{2}} \qquad (2.166)$$

$$\text{where} \qquad W_{AOFF} = W_A + V_{AOFF} \qquad (2.167)$$

and m is the conversion factor of the weighter having the dimension of volt^{-1}.

The signal $y_A'(t)$ is now combined with the main port output signal $y_M(t)$ by a two-way combiner to give the total output signal $y_T(t)$:

$$y_T(t) = \frac{y_A'(t) + y_M(t)}{\sqrt{2}} = \frac{1}{\sqrt{2}}\left[\frac{mW_{AOFF}\,y_A(t)}{\sqrt{2}} + y_M(t)\right] \qquad (2.168)$$

The signal $y_T(t)$ will now be divided into two equal parts by a two-way divider where one part will be applied at the second input port of the multiplier through a gain block with gain G and the other part will appear at the output port of the signal processing network as the final output signal $y_0(t)$ given by

$$y_0(t) = \frac{y_T(t)}{\sqrt{2}} = \frac{1}{2}\left[\frac{mW_{AOFF}\,y_A(t)}{\sqrt{2}} + y_M(t)\right] \qquad (2.169)$$

and will be delivered at the input port of the radio receiver.

2.10.3.2 Steady-State Value of the Weighter

The multiplier has a conversion factor p which has the dimension of volt^{-1}. It receives signals $[Ay_A(t)/\sqrt{2}]$ and $[Gy_0(t)]$ at its two input ports and converts $y_0(t)$ into its complex conjugate $y_0^*(t)$ before performing the multiplication.

As shown in Figure 2.4, the product signal at the output of the multiplier is passed through the gain block with gain $-g$, the RC integrating circuit, and a unity gain block, and finally applied at the control port of the weighter W_A.

Therefore, at the output of the multiplier, after the gain block $-g$, the product signal $y_{mo}(t)$ will be given by (2.23)

$$y_{mo}(t) = -\mu W_{AOFF} |y_A(t)|^2 - \frac{\sqrt{2}\mu |y_A(t) y_M^*(t)|}{m} \qquad (2.170)$$

$$\text{where} \qquad \mu = \frac{GAgpm}{4} \qquad (2.171)$$

and represents the total gain factor associated with the feedback circuit controlling the weighter. Since p and m are the conversion factors of the multiplier and the weighter, respectively, having the dimensions of volt^{-1}, μ will have the dimension of power^{-1}.

Again, with RC = T, the signal $y_{ii}(t)$ at the input port of the RC integrating circuit will be given by (2.25)

$$y_{ii}(t) = T\frac{dW_{AOFF}}{dt} + W_{AOFF} \qquad (2.172)$$

Hence, equating the two equations, (2.170) and (2.172), gives

$$\frac{dW_{AOFF}}{dt} = \frac{[1 + \mu |y_A(t)|^2]}{T} - \frac{\sqrt{2}\mu |y_A(t) y_M^*(t)|}{mT} \qquad (2.173)$$

Therefore, the steady-state value of the weighter W_{AOFF}, represented by W_{ASSO}, is given by (2.27)

$$W_{ASSO} = -V_{OFF} - \frac{\sqrt{2}\mu |y_A(t) y_M^*(t)|}{m[1 + \mu |y_A(t)|^2]} \qquad (2.174)$$

Substituting (2.165) into (2.174) gives

$$W_{ASSO} = -V_{OFF} - \frac{\sqrt{2}\mu |G_A(\theta_i) G_M(\theta_i)| p_i}{m[1 + \mu |G_A(\theta_i)|^2 p_i]} \qquad (2.175)$$

where $p_i = |f_i(t)|^2$ = power of the interference signal.

2.10.4 Final Output Power

From (2.169), the final output power p_o is given by

$$p_o = |y_o(t)|^2 = \frac{1}{4}\left[\frac{mW_{A\text{OFF}}\, y_A(t)}{\sqrt{2}} + y_M(t)\right]^2 \qquad (2.176)$$

$$\text{or } 4p_o = \frac{m^2[W_{A\text{OFF}}\, y_A(t)]^2}{2} + |y_M(t)|^2 + \frac{2m}{\sqrt{2}}\,\text{Re}\,W_{A\text{OFF}}\, y_A(t)y_M(t)$$

$$\qquad (2.177)$$

Substitution of (2.165) into (2.177) gives

$$4p_o = \frac{m^2 W_{A\text{OFF}}^2\,|G_A(\theta_i)|^2 p_i}{2} + |G_M(0)|^2 p_d + |G_M(\theta_i)|^2 p_i \qquad (2.178)$$

$$+ \frac{2mW_{A\text{OFF}}\,|G_A(\theta_i)G_M(\theta_i)|\,p_i}{\sqrt{2}}$$

Replacing $W_{A\text{OFF}}$ in (2.178) with the steady-state value W_{ASSO} in (2.175) and simplifying, the final steady-state output power p_{oSSO} is given by (Appendix G)

$$4p_{oSSO} = p_{Md} + \frac{p_{Mi}}{(1 + \mu p_{Ai})^2}\left[1 - \frac{(1 + \mu p_{Ai})mV_{A\text{OFF}}\sqrt{p_{Ai}}}{\sqrt{2}\sqrt{p_{Mi}}}\right]^2 \qquad (2.179a)$$

$$= p_{Md} + p_{\text{int}} = p_D + p_{\text{int}} \qquad (2.179b)$$

where

$p_{Md} = |G_M(0)|^2 p_d$ = power of the desired signal at the main output port;

$p_{Mi} = |G_M(\theta_i)|^2 p_i$ = power of the interference signal at the main output port;

$p_{Ai} = |G_A(\theta_i)|^2 p_i$ = power of the interference signal at the AUX-A output port;

$p_D = p_{Md}$ = the desired signal power in the final output power;

p_{int} = the interference signal power in the final output power.

From (2.179a), the interference signal power p_{int} is given by

$$p_{\text{int}} = \frac{p_{Mi}}{(1 + \mu p_{Ai})^2} \left[1 - \frac{(1 + \mu p_{Ai}) m V_{A\text{OFF}}}{\sqrt{2}} \frac{\sqrt{p_{Ai}}}{\sqrt{p_{Mi}}} \right]^2 \qquad (2.180a)$$

$$= \frac{p_{Mi}}{(1 + \mu p_{Ai})^2} \left[1 - \frac{(1 + \mu p_{Ai}) m V_{A\text{OFF}}}{\sqrt{2}} \frac{|G_A(\theta_i)|}{|G_A(\theta_i)|} \right]^2 \qquad (2.180b)$$

$$\text{where} \qquad \frac{\sqrt{p_{Ai}}}{\sqrt{p_{Mi}}} = \frac{|G_A(\theta_i)|}{|G_M(\theta_i)|} \qquad (2.181)$$

2.10.5 Null Depth

From (2.180), the null depth, corresponding to the given p_{int}, is therefore given by

$$\eta = \frac{p_{Mi}}{p_{\text{int}}} = \frac{[1 + \mu p_{Ai}]^2}{\left[1 - \dfrac{(1 + \mu p_{Ai}) m V_{A\text{OFF}}}{\sqrt{2}} \dfrac{|G_A(\theta_i)|}{|G_M(\theta_i)|} \right]^2}$$

$$= \frac{\eta_0}{\left[1 - \dfrac{(1 + \mu p_{Ai}) m V_{A\text{OFF}}}{\sqrt{2}} \dfrac{|G_A(\theta_i)|}{|G_M(\theta_i)|} \right]^2} \qquad (2.182)$$

$$\text{where} \qquad \eta_0 = [1 + \mu p_{Ai}]^2 \qquad (2.183)$$

is the null depth in the absence of any offset voltage (2.34a). Then,

$$\frac{\eta}{\eta_0} = \frac{1}{\left[1 - \dfrac{(1 + \mu p_{Ai}) m V_{A\text{OFF}}}{\sqrt{2}} \dfrac{|G_A(\theta_i)|}{|G_M(\theta_i)|} \right]^2} \qquad (2.184a)$$

$$= \frac{1}{(1 - K)^2} \qquad (2.184b)$$

$$\text{where} \qquad K = \frac{(1 + \mu p_{Ai}) m V_{A\text{OFF}}}{\sqrt{2}} \frac{|G_A(\theta_i)|}{|G_M(\theta_i)|} \qquad (2.185)$$

Example 2.15

The adaptive array system shown in Figure 2.4 utilizes a 4×4 beam former. The conversion factor m of the weighter W_A is $8\ V^{-1}$. The product of the loop gain μ and the AUX-A output power p_{Ai} has been found to be +35 dBm. The offset voltage associated with the weighter is 0.03 mV. Calculate the value of the ratio (η/η_0) if the location of the interference signal source is close to $+20°$ off the boresight so that the ratio of the gain functions $(|G_A(\theta_i)|/|G_M(\theta_i)|)$ is about 2.

Answer

Given

$$m = 8V^{-1}$$

$$\mu p_{Ai} = +35 \text{ dBm} = 3162$$

$$V_{A\text{OFF}} = 0.03 \text{ mV}$$

$$(|G_A(\theta_i)|/|G_M(\theta_i)|) = 2$$

Substitution of these values in (2.185) gives

$$K = 1.07356$$

Therefore, $(\eta/\eta_0) = 22.7$ dB

Thus, the null depth will display a value about 23 dB higher than the actual null depth that can be obtained with the adaptive array system being used, in the absence of any offset voltage.

Example 2.16

Calculate the values of the ratio (η/η_0) for different values of the weighter offset voltage for a few given values of the ratio (p_{Ai}/p_{Mi}).

Answer

Assuming $(p_{Ai}/p_{Mi}) = 1$, the offset voltage $V_{A\text{OFF}}$ is varied from 0.00 to 0.10 mV and the corresponding values of K are calculated. Then using (2.184b), the ratio (η/η_0) is obtained.

The calculations are repeated for

$$(|G_A(\theta_i)|^2/|G_M(\theta_i)|^2) = (p_{Ai}/p_{Mi}) = 2, 4 \text{ and } 10$$

The results are plotted in Figure 2.17. Note that the virtual null depth is created by the offset voltage over a certain voltage range and has very little or no effect outside this range. It is therefore advisable to avoid the regions of these offending offset voltages during a given measurement setup.

2.10.6 Feedback Loop with Phase Detector

A feedback loop whose multiplier has been ac coupled to the weighter may contain a phase detector as shown in Figure 2.18. The phase detector is dc coupled to the weighter and, hence is a potential source of offset voltage. In such situations, V_{AOFF} will represent the total modification to the weighter due to the combined effect.

2.11 Effect of Space Noise on System Performance

Any communication system is susceptible to noise whether it is the white noise generated by the circuit components or the space noise generated by celestial sources. In this section, the effect of space noise on system performance is discussed.

2.11.1 The Noise

In general, as far as the purity of the desired signal is concerned, any undesired signal could be regarded as a man-made noise generated by an undesired signal source situated at a particular location in space relative to the desired signal source. By contrast, space noise is a natural noise which is distributed uniformly throughout space and hence, irrespective of the incoming direction, the level of space noise is always the same. Therefore, every beam produced by a beam former, main or AUX, will see the same level of noise along its peak. This will appear at the corresponding output port, which has a normalized maximum gain of unity along the peak of the beam. In this section, an investigation is undertaken to determine the effect of this space noise on the performance of a given adaptive array system.

2.11.2 The Signals

Here again, for simplicity, the system of Figure 2.4 is considered where there are only two transmitting signal sources and where the associated beam former has only two active output ports.

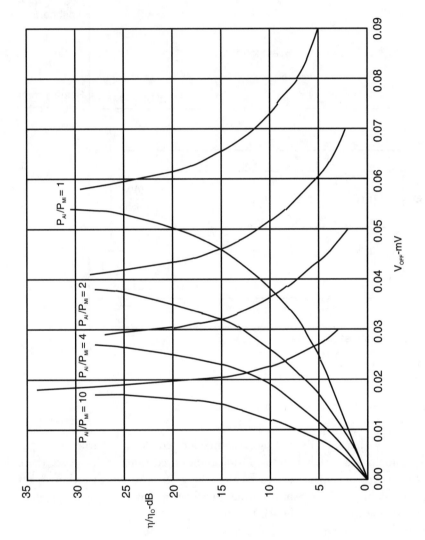

Figure 2.17 Variation of the ratio (η/η_O) with weighter offset voltage.

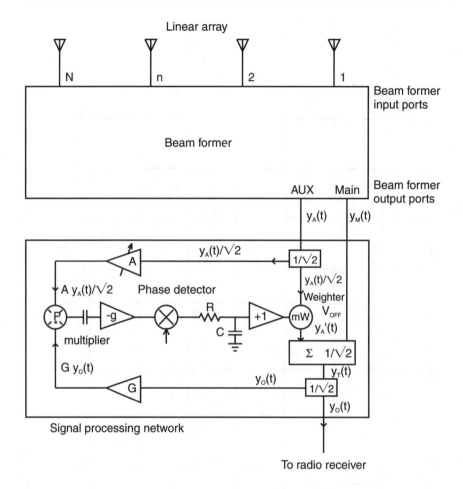

Figure 2.18 Feedback loop with a phase detector.

Thus, there is a desired signal source located along the array boresight or the main beam that is transmitting the desired signal $f_d(t)$. Also, there is an interference signal source located along an arbitrary direction θ_i off the boresight that is transmitting the interference signal $f_i(t)$ where θ_i does not coincide with any of the AUX beams. In addition, there is a noise source generating noise signal $f_n(t)$ uniformly distributed throughout space so that the noise level along every beam, main or AUX, is the same. It is assumed that the signals $f_d(t)$, $f_i(t)$, and $f_n(t)$ are uncorrelated.

Again, the beam former has only two active output ports, namely, the main port and an arbitrary AUX port designated the AUX-A port. All unused output ports of the beam former are match terminated.

Now, both the desired signal source and the space noise source are located along the main beam and hence the transfer function or gain between the main output port and the desired signal source and the space noise source will be designated $|G_M(0)|$ while that between the main output port and the interference signal source will be designated $|G_M(\theta_i)|$.

Likewise, the transfer function or gain between the AUX-A output port and the interference signal source will be designated $|G_A(\theta_i)|$.

Again, let the peak of the AUX-A beam be located along θ_A off the array boresight. Then, the transfer function or gain between the AUX-A output port and the noise source will be designated $|G_A(\theta_A)|$.

The signals $y_M(t)$ and $y_A(t)$ at the main and AUX-A output ports will therefore be given by

$$y_M(t) = |G_M(0)|f_d(t) + |G_M(\theta_i)|f_i(t) + |G_M(0)|f_n(t) \quad \text{(2.186a)}$$

$$y_A(t) = |G_A(\theta_i)|f_i(t) + |G_A(\theta_A)|f_n(t) \quad \text{(2.186b)}$$

These two signals are now delivered to the respective input ports of the signal processing network.

2.11.3 Signal Processing Network

Since the beam former has only two active output ports, the signal processing network will have only two input ports. One of the input ports is the main port and the other is an AUX port connected to a feedback loop. Thus, the signal processing network will contain only one weighter associated with the feedback loop. It is assumed that the weighter is automatic or self-adjusting.

2.11.3.1 Final Output Signal

As before, the signal $y_A(t)$ at the output of the AUX-A port will be divided into two equal parts $[y_A(t)/\sqrt{2}]$ where one part will be applied at the input port of a gain block with gain A and the other part will be applied at the input port of the weighter W_A. The signal $[Ay_A(t)/\sqrt{2}]$ at the output of gain block A will be applied at one of the two input ports of a multiplier while the signal $y_A'(t)$ at the output of the weighter will be applied at one of the two input ports of a two-way combiner.

The two-way combiner will now combine the signal $y_A'(t)$ from the output port of the weighter with the signal $y_M(t)$ from the main output port, to give the total output signal $y_T(t)$. The two signals, $y_A'(t)$ and $y_T(t)$, are

$$y_A'(t) = \frac{mW_A\,y_A(t)}{\sqrt{2}} \qquad (2.187)$$

$$y_T(t) = \frac{y_A'(t) + y_M(t)}{\sqrt{2}} = \frac{1}{\sqrt{2}}\left[\frac{mW_A\,y_A(t)}{\sqrt{2}} + y_M(t)\right] \qquad (2.188)$$

Again, the total output signal $y_T(t)$ is divided into two equal parts where one part is applied to the second input port of the multiplier through a gain block with gain G while the other part appears at the output port of the signal processing network as the final output signal $y_o(t)$ given by

$$y_o(t) = \frac{y_T(t)}{\sqrt{2}} = \frac{1}{2}\left[\frac{mW_A\,y_A(t)}{\sqrt{2}} + y_M(t)\right] \qquad (2.189)$$

Note that in (2.187) m is the conversion factor of the weighter having the dimension of volt^{-1}. The final output signal $y_o(t)$ will now be delivered to the input port of the radio receiver.

2.11.3.2 Steady-State Value of the Weighter

As has been mentioned in previous sections, the multiplier has a conversion factor p which has a dimension of volt^{-1}. It receives signals $[Ay_A(t)/\sqrt{2}]$ and $[Gy_o(t)]$ at its two input ports. It converts $y_o(t)$ into its complex conjugate $y_o^*(t)$ and performs the multiplication.

The resulting product signal at the output of the multiplier will pass through a gain block $-g$, an RC integrator circuit, a unity gain block, and finally be applied at the control input port of the weighter.

The multiplier output signal $y_{mo}(t)$ after the gain block $-g$ and the signal $y_{ii}(t)$ at the input port of the RC integrator are given by (2.23 and 2.25)

$$y_{mo}(t) = -\mu\left[W_A\,|y_A(t)|^2 - \frac{\sqrt{2}\,y_A(t)\,y_M^*(t)}{m}\right] \qquad (2.190)$$

$$y_{ii}(t) = T\frac{dW_A}{dt} + W_A \qquad (2.191)$$

where T = RC and

$$\mu = \frac{GAgpm}{4} \tag{2.192}$$

which represents the total gain factor associated with the feedback circuit controlling the weighter (2.24) and has the dimension of power^{-1}.

Again, since the signals $y_{mo}(t)$ and $y_{ii}(t)$ are the same, (2.190) and (2.191) give

$$\frac{dW_A}{dt} = \frac{1 + \mu|y_A(t)|^2 W_A}{T} - \frac{\sqrt{2}\mu|y_A(t)y_M^*(t)|}{mT} \tag{2.193}$$

At steady state $(dW_A/dt) = 0$, and hence (2.193) gives

$$W_{ASS} = -\frac{\sqrt{2}\mu|y_A(t)y_M^*(t)|}{m(1 + \mu|y_A(t)|^2)} \tag{2.194}$$

Substituting (2.186) into (2.194) gives

$$W_{ASS} = -\frac{\sqrt{2}\mu|G_A(\theta_i)G_M(\theta_i)|p_i + |G_A(\theta_A)G_M(0)|p_n}{m[1 + \mu|G_A(\theta_A)|^2 p_n + \mu|G_A(\theta_i)|^2 p_i]} \tag{2.195}$$

where $p_i = |f_i(t)|^2$ = the interference signal power and $p_n = |f_n(t)|^2$ = the average noise power.

2.11.4 Final Output Power

From (2.189), the final output power p_o will be given by

$$p_o = \frac{1}{4}\left[\frac{mW_A y_A(t)}{\sqrt{2}} + y_M(t)\right]^2 \tag{2.196a}$$

or $\quad 4p_0 = \dfrac{m^2[W_A y_A(t)]^2}{2} + |y_M(t)|^2 + \dfrac{2m\,\mathrm{Re}\,W_A[y_A(t)y_M(t)]}{\sqrt{2}}$

$$\tag{2.196b}$$

Substituting (2.186) into (2.196b) gives

$$4p_o = \frac{m^2 W_A^2 [|G_A(\theta_i)|^2 p_i + |G_A(\theta_A)|^2 p_n]}{2}$$

$$+ [|G_M(0)|^2 p_d + |G_M(\theta_i)|^2 p_i + |G_M(0)|^2 p_n] \qquad (2.197)$$

$$+ \frac{2m W_A [|G_A(\theta_i) G_M(\theta_i)| p_i + |G_A(\theta_A) G_M(0)| p_i]}{\sqrt{2}}$$

where $p_d = |f_d(t)|^2$ = the desired signal power. Replacing W_A in (2.197) by W_{ASS} from (2.195) and simplifying, the final steady-state output power p_{oSSn} is given by (Appendix H)

$$4p_{oSSn} = p_{Md} + p_{Mi} \left[1 - \frac{\mu p_{Ai} + \mu K_{AIA} p_{An}}{1 + \mu p_{Ai} + \mu p_{An}} \right]^2$$

$$+ p_{Mn} \left[1 - \frac{\mu K_{AAI} p_{Ai} + \mu p_{An}}{1 + \mu p_{Ai} + \mu p_{An}} \right]^2 \qquad (2.198a)$$

$$= p_{Md} + p_{Mi} \frac{[1 + \mu p_{An}(1 - K_{AIA})]^2}{[1 + \mu p_{Ai} + \mu p_{An}]^2}$$

$$+ p_{Mn} \frac{[1 + \mu p_{Ai}(1 - K_{AAI})]^2}{[1 + \mu p_{Ai} + \mu p_{An}]^2} \qquad (2.198b)$$

$$= p_D + p_{\text{int}} + p_{no}$$

where

$p_{Md} = |G_M(0)|^2 p_d$ = the desired signal power at the main output port;

$p_{Mi} = |G_M(\theta_i)|^2 p_i$ = the interference signal power at the main output port;

$p_{Mn} = |G_M(0)|^2 p_n$ = the average noise power at the main output port;

$p_{Ai} = |G_A(\theta_i)|^2 p_i$ = the interference signal power at the AUX-A output port;

$p_{An} = |G_A(\theta_A)|^2 p_n$ = the average noise power at the AUX-A output port;

$$K_{AIA} = \frac{1}{K_{AAI}} = \frac{|G_A(\theta_i)/|G_A(\theta_A)|}{|G_M(\theta_i)/|G_M(0)|} = \frac{|G_A(\theta_i)|}{|G_M(\theta_i)|}$$

where the normalized gain along the peak of a beam has been regarded as unity. The three power components present in the final output power p_{oSSn}, appearing at the output port of the signal processing network, are given by

$$p_D = p_{Md} = \text{the desired signal power} \qquad (2.199a)$$

$$p_{\text{int}} = p_{Mi} \frac{[1 + \mu p_{An}(1 - K_{AIA})]^2}{[1 + \mu p_{Ai} + \mu p_{An}]^2}$$

$$= \text{the interference signal power} \qquad (2.199b)$$

$$p_{no} = p_{Mn} \frac{[1 + \mu p_{Ai}(1 - K_{AAI})]^2}{[1 + \mu p_{Ai} + \mu p_{An}]^2}$$

$$= \text{the average noise power} \qquad (2.199c)$$

2.11.5 Null Depth

The expected null depth corresponding to p_{int} given in (2.199b) is thus

$$\eta = \frac{p_{Mi}}{p_{\text{int}}} = \frac{[1 + \mu p_{Ai} + \mu p_{An}]^2}{[1 + \mu p_{An}(1 - K_{AIA})]^2}$$

$$= (1 + \mu p_{Ai})^2 \frac{[1 + \mu p_{Ai} + \mu p_{An}]^2}{[(1 + \mu p_{Ai})\{1 + \mu p_{An}(1 - K_{AIA})\}]^2} \qquad (2.200a)$$

$$= \eta_0 \frac{[1 + \mu p_{Ai} + \mu p_{An}]^2}{[(1 + \mu p_{Ai})\{1 + \mu p_{An}(1 - K_{AIA})\}]^2}$$

$$= \eta_0 F_{nn}$$

where η_0 is the null depth with the absence of the noise source and F_{nn} is the null modification factor due to noise. Then

$$\frac{\eta}{\eta_0} = F_{nn} = \frac{[1 + \mu p_{Ai} + \mu p_{An}]^2}{[(1 + \mu p_{Ai})\{1 + \mu p_{An}(1 - K_{AIA})\}]^2} \qquad (2.200b)$$

and is equal to unity in the absence of the space noise.

For $\mu p_{An} \ll p_{Ai}$, (2.200b) reduces to

$$\frac{\eta}{\eta_0} = F_{nn} = \frac{1}{[\{1 + \mu p_{An}(1 - K_{AIA})\}]^2} \tag{2.201}$$

2.11.6 Final Output Noise

Again, from (2.199c), the total noise power present at the output port of the signal processing network is

$$p_{no} = p_{Mn} \frac{[1 + \mu p_{Ai}(1 - K_{AAI})]^2}{[1 + \mu p_{Ai} + \mu p_{An}]^2} \tag{2.202}$$

$$= p_{Mn} F_{Mn}$$

where, as mentioned above, p_{Mn} is the noise power contained in the main port output signal and F_{Mn} is the main port noise modification factor caused by the feedback loop and is given by

$$F_{Mn} = \frac{[1 + \mu p_{Ai}(1 - K_{AAI})]^2}{[1 + \mu p_{Ai} + \mu p_{An}]^2} \tag{2.203}$$

Thus, the noise picked up by the main beam and appearing at the main output port will be modified by the factor F_{Mn} before being delivered to the radio receiver. With the feedback loop disengaged, F_{Mn} is unity and the noise contained in the total output power is just the noise present in the main output port signal.

For $\mu p_{An} << \mu p_{Ai}$, (2.203) reduces to

$$F_{Mn} = \frac{[1 + \mu p_{Ai} - \mu p_{Ai} K_{AAI}]^2}{[1 + \mu p_{Ai}]^2} \tag{2.204}$$

$$= \left[1 - \frac{(\mu p_{Ai} K_{AAI})}{1 + \mu p_{Ai}}\right]^2$$

For $\mu p_{Ai} >> 1$, (2.204) reduces to

$$F_{Mn} = (1 - K_{AAI})^2 \tag{2.205}$$

2.11.6.1 Special Case

Consider the special case where the location of the interference signal source is such that the transfer function or gain between the interference signal source and the main output port is equal to that between the interference signal source and the AUX-A output port; $|G_M(\theta_i)| = |G_A(\theta_i)|$. Then

$$K_{AIA} = K_{AAI} = 1 \qquad (2.206)$$

Hence, (2.199b) and (2.199c) reduce to

$$p_{\text{int}} = \frac{p_{Mi}}{[1 + \mu p_{Ai} + \mu p_{An}]^2} \qquad (2.207a)$$

$$p_{no} = \frac{p_{Mn}}{[1 + \mu p_{Ai} + \mu p_{An}]^2} \qquad (2.207b)$$

Therefore, the corresponding null depth will become

$$\eta = [1 + \mu p_{Ai} + \mu p_{An}]^2 \qquad (2.208)$$

and the noise modification factor will become the inverse of the null depth

$$F_{Mn} = \frac{1}{[1 + \mu p_{Ai} + \mu p_{An}]^2} \qquad (2.209)$$

Example 2.17

An adaptive array system utilizes a 4×4 beam former. Assuming beam-(+1) is beam A, estimate the values of the modification factors F_{nn} and F_{Mn} for locations of the interference signal source relative to $+14°$ where $|G_M(\theta_i)| = |G_A(\theta_i)|$.

Answer

From Figure 1.6, we can see that if the location θ_i of the interference signal source is greater than $+14°$ then $|G_A(\theta_i)|$ is larger than $|G_M(\theta_i)|$ and hence K_{A1A} is greater than one, and K_{AA1} is less than one. On the other hand, if θ_i is smaller than $+14°$, then $|G_A(\theta_i)|$ is smaller than $|G_M(\theta_i)|$ and hence K_{A1A} is smaller than one and K_{AA1} is larger than one.

Case 1

$K_{A1A} > 1$. Hence, in (2.202), $(1 - K_{A1A})$ is negative, which will make

$$1 + \mu p_{An}(1 - K_{A1A}) < 1$$

Therefore, F_{nn} is greater than one, causing the null depth to increase. Again, $K_{AA1} < 1$. Hence, in (2.205), $(1 - K_{AA1})$ is less than one and so

$$(1 - K_{AA1})^2 < 1$$

Therefore, F_{nn} is less than one, which will cause the main port noise to decrease.

Case 2

$K_{A1A} < 1$. Hence, from (2.202), $(1 - K_{A1A})$ is positive, which will make

$$1 + \mu p_{An}(1 - K_{A1A}) > 1$$

Therefore, F_{nn} will become less than one causing the null depth to decrease. Again, $K_{AA1} < 1$. Hence, in (2.205), $(1 - K_{AA1})$ is larger than one and so

$$(1 - K_{AA1})^2 > 1$$

Therefore, F_{nn} will become larger than one, which will cause the main port noise to increase.

2.12 Feedback Loop

From the discussions presented in the foregoing sections, it becomes apparent that in order to remove the interference signals from the desired signal, the signal processing network will have to use a feedback loop in association with a weighter. In essence, the feedback loop is responsible for adjusting the weighter parameters so that the corresponding AUX port signal will acquire the proper amplitude and phase to cancel the interference signal from the total output signal so that the final output signal at the output port of the signal processing network will be free of the interference signal.

2.12.1 Stability of the Feedback Loop

Since the feedback loop is one of the key elements for the removal of interference signals, it is important for efficient performance of the signal processing network that the feedback loop remain unconditionally stable throughout the operation.

2.12.2 Closed-Loop Transfer Function of a Feedback Loop

To examine the stability of a feedback loop, a single-loop system, given by Figure 2.4 in Section 2.2, is considered that has only two transmitting signal sources and the associated beam former has only two active output ports.

However, to analyze the stability of a feedback loop, it is necessary first to derive the closed-loop transfer function of the desired feedback loop.

2.12.2.1 Loop Differential Equation

From (2.26), the time derivative of the weighter W_A, located inside the feedback loop connected to the AUX-A output port, is given by

$$
\begin{aligned}
\frac{dW_A}{dt} &= -\frac{[1 + \mu|y_A(t)|^2]}{T} W_A - \frac{\sqrt{2}\mu|y_A(t)y_M^*(t)|}{mT} \\
&= -\frac{[1 + \mu|G_A(\theta_i)|^2 p_i]}{T} W_A - \frac{\sqrt{2}\mu|G_A(\theta_i)G_M(\theta_i)|p_i}{mT} \quad (2.210) \\
&= -\frac{(1 + \mu p_{Ai})}{T} W_A - \frac{\sqrt{2}\mu q_{MAi}}{mT}
\end{aligned}
$$

$$
\text{where} \qquad |y_A(t)|^2 = |G_A(\theta_i)|^2 p_i = p_{Ai} \qquad (2.211a)
$$

$$
\text{and} \qquad |y_A(t)y_M^*(t)| = |G_A(\theta_i)G_M(\theta_i)|p_i = q_{MAi} \qquad (2.211b)
$$

where p_{Ai} = the total interference signal power at the AUX-A output port, and q_{MAi} = the cross-gain interference signal power between the main output port and the AUX-A output port.

Also, as given by (2.24),

$$
\mu = \frac{GAgmp}{4}
$$

and represents the total gain factor associated with the feedback circuit controlling the weighter.

Note that p and m are the conversion factors of the multiplier and the weighter, respectively (Figure 2.4), having the dimensions of volt^{-1} and hence μ has the dimension of power^{-1}. Following (2.72b), let

$$\mu' = \mu/pm$$

where, as stated in (2.72b), μ' is a dimensionless gain factor. Then (2.210) can be written

$$\frac{dW_A}{dt} = -\frac{(1 + \mu p_{Ai})}{T} W_A - \frac{\sqrt{2}\mu' pq_{MAi}}{T} \qquad (2.212)$$

Put

$$pq_{MAi} = v_{MAi} \qquad (2.213)$$

where v_{MAi} has the dimension of volt. Then (2.212) becomes

$$\frac{dW_A}{dt} = -\frac{(1 + \mu p_{Ai})}{T} W_A - \frac{\sqrt{2}\mu' v_{MAi}}{T} \qquad (2.214)$$

2.12.2.2 Loop Transfer Function

To calculate the loop transfer function, (2.214) has to be transformed from the time domain into the complex frequency domain $s = \sigma + j\omega$. Then (2.214) gives

$$sW_A = -\frac{(1 + \mu p_{Ai})}{T} W_A - \frac{\sqrt{2}\mu' v_{MAi}}{T}$$

or $\qquad \left(s + \frac{1 + \mu p_{Ai}}{T}\right) W_A = -\frac{\sqrt{2}\mu' v_{MAi}}{T} \qquad (2.215)$

Hence, the closed-loop transfer function H_{Ai} of the feedback loop connected to the AUX-A output port, due to the interference signal $f_i(t)$, is given by

$$H_{Ai} = \frac{W_A}{v_{MAi}} = -\frac{\sqrt{2}\mu'/T}{s + \dfrac{1 + \mu p_{Ai}}{T}} = -\frac{\sqrt{2}\mu'/T}{s + s_{Ai}} \qquad (2.216)$$

where $\qquad s_{Ai} = \frac{1 + \mu p_{Ai}}{T} \qquad (2.217)$

Alternate Expression for Loop Transfer Function

Equation (2.216) can also be expressed as

$$H_{Ai} = -\frac{\sqrt{2}\mu'}{1 + \mu p_{Ai}} \frac{(1 + \mu p_{Ai})/T}{s + \frac{1 + \mu p_{Ai}}{T}} = -K\frac{\omega_1}{s + \omega_1} \qquad (2.218)$$

$$\text{where} \qquad K = \frac{\sqrt{2}\mu'}{1 + \mu p_{Ai}} \qquad (2.219a)$$

$$\text{and} \qquad \omega_1 = \frac{1 + \mu p_{Ai}}{T} \qquad (2.219b)$$

2.12.2.3 The Root Locus and the Bode Plot of the Loop Transfer Function

Equation (2.216) can be used to plot the root locus of the transfer function while (2.218) is suitable for Bode plot.

Figure 2.19(a) shows a plot of root locus in the complex frequency plane of $s = \sigma + j\omega$. The transfer function is seen to have only one pole at

$$s_{Ai} = \sigma_{Ai} = -(1 + \mu p_{Ai})/T$$

With the feedback loop disengaged, $\mu p_{Ai} = 0$, and the pole is located at $s_{Ai} = \sigma_{Ai} = -(1/T)$. On the other hand, with the feedback loop engaged, as μp_{Ai} increases, σ_{Ai} increases and, hence, the pole moves toward negative infinity. Thus, the pole could not possibly cross into the right half section of the complex frequency plane and, therefore the system will remain unconditionally stable.

The Bode plot of the transfer function is shown in Figure 2.19(b). We can see that the gain remains constant at K up to the frequency $\omega = \omega_1$. With further increase in frequency ω, the gain starts to fall off at 6 dB/octave. Similarly, the phase remains at 180° up to the frequency $\omega = \omega_1$, at which point it drops to 90° and remains there with further increase in frequency ω.

2.12.3 Feedback Loop and the Convergence Time

The time it takes a weighter to reach the steady state is known as the *convergence time* of the weighter. The effectiveness of some signal processing

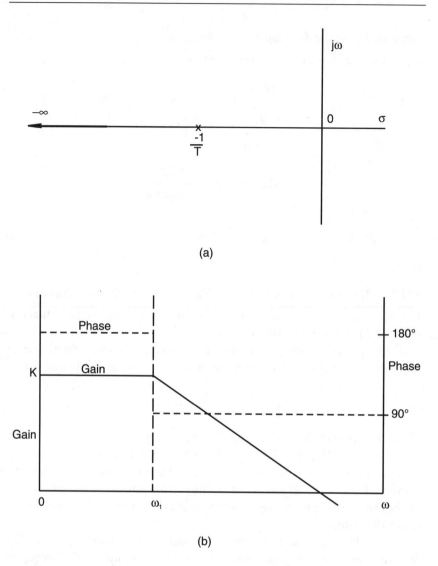

Figure 2.19 (a) Root locus and (b) Bode plot of a single-pole feedback loop.

operations, such as those utilizing the LMS algorithm, could be affected by the convergence time.

2.12.3.1 Convergence Time as a Function of Loop Gain

From (2.216) we can see that the transfer function H_{Ai} and hence the weighter W_A changes with time in accordance with $e^{-(\sigma Ai)t}$. Thus, the convergence time will depend on σ_{Ai} and, hence, according to (2.217), on

μp_{Ai}. However, as has been mentioned in Section 2.2.4, μp_{Ai} is the open-loop gain of the feedback loop. Thus, the convergence time is a function of the open-loop gain and hence will increase or decrease with increasing or decreasing loop gain.

The gain blocks A and G are usually equipped with AGC and, hence, under normal operation the loop gain will remain constant over a fairly wide range of p_{Ai}. In a multisource system, however, a given loop will contain different signals with different strengths. Since the gain of an AGC will be dictated by the strongest signal, the weaker signals will experience smaller loop gains and, consequently, smaller convergence times.

The wide difference in convergence times associated with various signals involved may cause some undesirable effects. The very fast converging signals will make the weighting coefficients jittery, causing large variation in output power. On the other hand, very slow converging signals take so long to converge that the weighter will permit a considerable amount of interference signal into the receiver before reaching the steady state. To prevent this, the dependence of the pole on the loop gain should be avoided or severely restricted.

2.12.3.2 Modified Feedback Loop

As seen from the above, both very fast and very slow convergence times, associated with a weighter, are equally harmful. However, with a few modifications, it is possible to obtain the correct convergence time from a given feedback loop.

Principle of Loop Modification

From circuit theory, we know that for an even number of poles, there cannot be a root locus on the left side of the poles. Therefore, if a second pole is introduced into the feedback loop then, with increasing loop gain, the two poles would simply move toward each other and coalesce into a single pole at a point in between.

A modification to the basic feedback loop of Figure 2.4 is shown in Figure 2.20, where a second RC circuit has been placed between the multiplier and the gain block $-g$. For identification, the original RC pair is designated $R_2 C_2$ and the new pair $R_1 C_1$.

Poles of the Modified Feedback Loop

Let an open circle (O) be a point between the multiplier and the $R_1 C_1$ pair (Figure 2.20). Then looking toward the multiplier from O, the signal $y_{om1}(t)$ is given by (2.23)

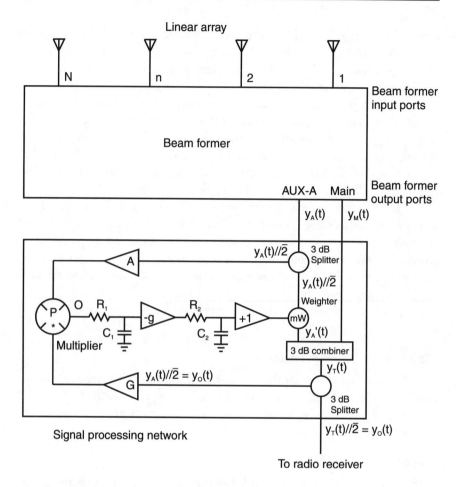

Figure 2.20 Basic feedback loop modified with a second pole.

$$y_{om1}(t) = \mu_1 |y_A(t)|^2 W_A + \frac{\sqrt{2}\mu_1 |y_A(t) y_M^*(t)|}{m} \qquad (2.220)$$

where from (2.24)

$$\mu_1 = \frac{\mu}{g} = \frac{GApm}{4} \qquad (2.221)$$

Similarly, looking toward the $R_1 C_1$ pair from the point O the signal $y_{ii1}(t)$ is given by [Figure 2.21 and Appendix K]

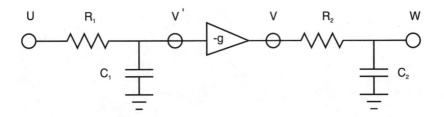

Figure 2.21 Signal at the output port of the multiplier in a two-pole feedback loop [$u = y_{om1}(t)$].

$$y_{ii1}(t) = -\frac{T_1 T_2}{g} \frac{d^2 W_A}{dt^2} - \frac{T_1 + T_2}{T_1 T_2} \frac{dW_A}{dt} - \frac{W_A}{g} \qquad (2.222)$$

Combining (2.220) and (2.222) gives

$$\frac{d^2 W_A}{dt^2} + \frac{T_1 + T_2}{T_1 T_2} \frac{dW_A}{dt} + \frac{1 + \mu |y_A(t)|^2]}{T_1 T_2} W_A = -\frac{\sqrt{2}\mu' p |y_A(t) y_M^*(t)|}{T_1 T_2}$$

or $$\frac{d^2 W_A}{dt^2} + \frac{T_1 + T_2}{T_1 T_2} \frac{dW_A}{dt} + \frac{1 + \mu p_{Ai}}{T_1 T_2} W_A = -\frac{\sqrt{2}\mu' v_{MAi}}{T_1 T_2} \qquad (2.223)$$

where as before, $\mu' = \mu/pm$ and

$$|y_A(t)|^2 = |G_A(\theta_i)|^2 p_i = p_{Ai}$$
$$|y_A(t) y_M^*(t)| = |G_A(\theta_i) G_M(\theta_i)| p_i = q_{MAi}$$
$$pq_{MAi} = v_{MAi}$$

As mentioned earlier, q_{Mai} has the dimension of power and v_{MAi} has the dimension of volt. Therefore, in the complex frequency domain, (2.223) gives

$$W_A \left[s^2 + \frac{T_1 + T_2}{T_1 T_2} s + \frac{1 + \mu p_{Ai}}{T_1 T_2} \right] = -\frac{\sqrt{2}\mu' v_{MAi}}{T_1 T_2} \qquad (2.224)$$

Hence, the corresponding closed loop transfer function H_{Ai} of the feedback loop becomes

$$H_{Ai} = -\frac{\sqrt{2}\mu'/(T_1T_2)}{s^2 + \frac{T_1 + T_2}{T_1T_2}s + \frac{1 + \mu p_{Ai}}{T_1T_2}} \qquad (2.225)$$

The poles of (2.225) are given by the expression

$$s^2 + \frac{T_1 + T_2}{T_1T_2}s + \frac{1 + \mu p_{Ai}}{T_1T_2} = 0 \qquad (2.226)$$

Equation (2.226) has two roots given by

$$s_{p1,p2} = -\frac{T_1 + T_2}{2T_1T_2} \pm \frac{T_1 + T_2}{2T_1T_2}\sqrt{1 - \frac{4T_1T_2(1 + \mu p_{Ai})}{(T_1 + T_2)^2}} \qquad (2.227)$$

If the two RC pairs are chosen such that $T_2 \gg T_1$, then (2.227) will reduce to

$$s_{p1,p2} = -\frac{1}{2T_1} \pm \frac{1}{2T_1}\sqrt{1 - \frac{1 + \mu p_{Ai}}{T_2/4T_1}} \qquad (2.228)$$

Example 2.18

Draw the root locus of the two poles s_{p1} and s_{p2} given by (2.228).

Answer

To obtain the root locus, μp_{Ai} in (2.228) is varied from 0 to $\gg(T_2/4T_1)$.

a) $1 + \mu p_{Ai} = 1$ or $\mu p_{Ai} = 0$ (feedback loop deactivated)

$$\begin{aligned} s_{p1,p2} &= -\frac{1}{2T_1} \pm \frac{1}{2T_1}\sqrt{1 - \frac{4T_1}{T_2}} \\ &= -\frac{1}{2T_1} \pm \frac{1}{2T_1}\left[1 - \frac{2T_1}{T_2}\right] \qquad (2.229) \\ &= -\frac{1}{2T_1} \pm \left[\frac{1}{2T_1} - \frac{1}{T_2}\right] \end{aligned}$$

Therefore, $\quad s_{p1} = -\frac{1}{T_1} + \frac{1}{T_2} \sim -\frac{1}{T_1} \qquad (2.230a)$

$$s_{p2} = -\frac{1}{T_2} \qquad (2.230b)$$

Note that since $T_2 \gg T_1$, the pole s_{p1} will be dominant and determine the weighter convergence time.

b) $1 < 1 + \mu p_{Ai} < T_2/4T_1$

$$s_{p1,p2} = -\frac{1}{2T_1} \pm \frac{1}{2T_1}\left[1 - \frac{2T_1}{T_2}(1 + \mu p_{Ai})\right] \qquad (2.231)$$

$$= -\frac{1}{2T_1} \pm \left[\frac{1}{2T_1} - \frac{(1 + \mu p_{Ai})}{T_2}\right]$$

Therefore, $\quad s_{p1} = -\frac{1}{T_1} + \frac{(1 + \mu p_{Ai})}{T_2} \qquad (2.232a)$

$$s_{p2} = -\frac{(1 + \mu p_{Ai})}{T_2} \qquad (2.232b)$$

Thus, s_{p1} will move toward right and s_{p2} will move toward left by the same distance given by $(1 + \mu p_{Ai})/T_2$.

c) $(1 + \mu p_{Ai}) = T_2/4T_1$

$$(1 + \mu p_{Ai}) = T_2/4T_1 \qquad (2.233)$$

$$s_{p1,p2} = -\frac{1}{2T_1}$$

Therefore, $\quad s_{p1} = -\frac{1}{2T_1} \qquad (2.234a)$

$$s_{p2} = -\frac{1}{2T_1} \qquad (2.234b)$$

At this point the two poles will coalesce into a single pole.

d) $(1 + \mu p_{Ai}) \gg T_2/4T_1$

$$s_{p1,p2} = -\frac{1}{2T_1} \pm \frac{1}{2T_1}\sqrt{-\frac{4T_1}{T_2}(1 + \mu p_{Ai})} \qquad (2.235a)$$

$$= -\frac{1}{2T_1} \pm j\sqrt{\frac{1 + \mu p_{Ai}}{T_1 T_2}} \qquad (2.235b)$$

$$= -\frac{1}{2T_1} \pm j\sqrt{F} \qquad (2.235c)$$

$$= -\frac{1}{2T_1} \pm j\omega_n$$

$$\text{where} \qquad \omega_n = \sqrt{F} = \sqrt{\frac{1 + \mu p_{Ai}}{T_1 T_2}} \qquad (2.236)$$

and is known as the natural frequency of the poles. Therefore,

$$s_{p1} = -\frac{1}{2T_1} - j\omega_n \qquad (2.237a)$$

$$s_{p2} = -\frac{1}{2T_1} + j\omega_n \qquad (2.237b)$$

In this case the two poles appear to display decayed oscillation where the angular frequency of oscillation is ω_{Ai} and the time constant of the decay is $2T_1$. Thus, although ω_{Ai} will vary with the loop gain μp_{Ai}, the convergence time, given by the real part of the poles, will remain constant irrespective of the value of the loop gain.

The complete root locus is shown in Figure 2.22.

Example

A single-loop adaptive system has an open-loop gain of 30 dB. The two RC pairs in the feedback loop are so chosen that T_1 is 60 μs and T_2 is 600 μs. Calculate the natural frequency of oscillation of the poles of the feedback loop.

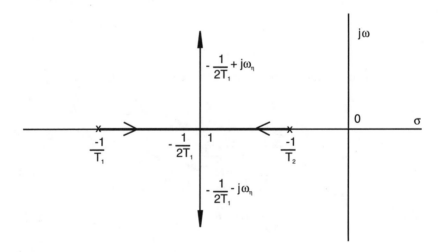

Figure 2.22 Root locus of a two-pole feedback loop.

Answer

$$\text{Here,} \quad \mu p_{Ai} = 30 \text{ dB}$$

$$T_1 = 60 \times 10^{-6} \text{ sec}$$

$$T_2 = 600 \times 10^{-6} \text{ sec}$$

Therefore, from (2.236),

$$\omega_n^2 = 10^6/36$$

$$\text{or} \quad \omega_n = 167 \text{ rad/s}$$

$$\text{or} \quad f_n = 26.5 \text{ Hz}$$

Effect of Natural Frequency on System Performance

From (2.236), we can see that the natural frequency of the feedback loop increases with increasing loop gain μp_{Ai}. Thus, there is a possibility that at some values of the loop gain, the natural frequency f_n will reach a value comparable with a frequency belonging to the signal frequency band which is passing through the weighter. In that case, the two frequencies may correlate to cause a higher misadjustment error as well as a reduction in stability margin. However, the frequency band passing through the weighter is the signal intermediate frequency (IF), which is usually of the order of 10 MHz. On the other hand, a realistic value of f_n is of the order of 30 Hz and hence a correlation is not very likely.

Again, a frequency of 26.5 Hz corresponds to a period of about 38×10^{-3} sec. Thus, if $T_2 = 0.6 \times 10^{-3}$ sec, then the weighter will reach steady state within a fraction of the period of oscillation. Consequently, the weighter will essentially see a dc of negative or positive polarity and never a full cycle of oscillation.

2.12.3.3 High-Frequency Poles

Real devices are mostly band limited and hence at higher frequencies will possess poles located on the left half of the complex s plane. These poles will cause the root locus of the two-pole circuit to bend toward the imaginary axis. It is thus possible that at some values of the loop gain μp_{Ai}, the root locus will cross the imaginary line and enter the right half of the s plane. This will obviously cause instability.

The root locus of a two-pole feedback loop with high-frequency poles is shown in Figure 2.23, where p is the total number of poles. From the triangle made by the real axis, the imaginary axis and the root locus,

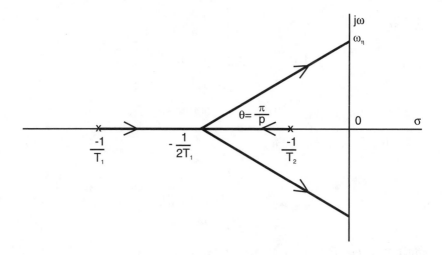

Figure 2.23 Effect of high-frequency poles on the root locus of a two-pole feedback loop.

$$\tan \theta = 2 T_1 \omega_n$$

$$\text{or} \quad \omega_n = (\tan \theta)/2 T_1 \quad\quad (2.238)$$

For a two-pole system, $p = 2$. Hence, from the root locus of a two-pole system shown in Figure 2.22, θ is found to be

$$\theta = 90° = \pi/2 \quad\quad (2.239)$$

Therefore, for a p-pole system, θ will be given by

$$\theta = \pi/p \qu\quad (2.240)$$

Example 2.19

A two-pole feedback loop has two high-frequency poles. The two RC pairs have been so chosen that $T_1 = 60 \ \mu s$ and $T_2 = 600 \ \mu s$. Calculate the loop gain required for the root locus to cross the imaginary axis.

Answer

For $p = 4$ $\theta = \pi/4 = 45°$. Hence, for $T_1 = 60 \ \mu s$, (2.238) gives

$$\omega_n = 8.333 \times 10^{-3} \ \text{rad/s}$$

Therefore, for $T_2 = 600 \ \mu s$, (2.236) gives

$$\mu p_{Ai} = 64 \text{ dB}$$

Note that a loop gain of the order of 64 dB is not a very realistic value in practice.

Example 2.20

A two-pole feedback loop has a loop gain of 30 dB. The two RC pairs have been so chosen that $T_1 = 60 \ \mu s$ and $T_2 = 600 \ \mu s$. Calculate the number of poles required for the root locus to cross the imaginary axis.

Answer

From the given values of μp_{Ai}, T_1, and T_2, the natural frequency ω_n is obtained from (2.236) as

$$\omega_n = 167 \text{ rad/s}$$

Therefore, from (2.238), the value of θ corresponding to this ω_n will be given by $\theta = 1.15°$, and, hence, from (2.239), the number of poles required to produce this value of θ is $p = 157$. Note that it is not very likely that a single feedback loop will contain that many high-frequency poles.

2.12.3.4 Nonoscillatory Two-Pole Feedback Loop

Note, however, that it is not difficult to build a nonoscillatory feedback loop. A few additional circuit components added to a two-pole feedback loop could make it unconditionally stable [1].

From (2.236), we can see that the roots of a quadratic equation become complex when F, the quantity under the square root or having power half, becomes negative. Thus,

$$F = 1 - \frac{1 + \mu p_{Ai}}{T_2/4T_1} = \text{negative} \qquad (2.241)$$

This could be prevented if in (2.241), $1 + \mu p_{Ai}$ could be placed in the denominator; that is, if

$$F = 1 - \frac{T_2/4T_1}{1 + \mu p_{Ai}} \qquad (2.242)$$

Hence, if $T_2 \gg 4T_1$, then F will always be positive irrespective of the value of μp_{Ai}. However, from the properties of a quadratic equation, this can only happen if the coefficient of dW_A/dt in (2.223) or that of s in (2.226) contains $(1 + \mu p_{Ai})$.

Now, in (2.223), the coefficient of W_A contains $(1 + \mu p_{Ai})$. Therefore, in Figure 2.20, if a time derivative of W_A is added in the feedback line before the gain block G, then its coefficient will also contain the term $(1 + \mu p_{Ai})$.

One way to achieve this is shown in Figure 2.24. Here, a second weighter with input $T_2(dW_A/dt)$ is placed in parallel to the first one. Then, at the input port of the gain block G, the signal will be

$$y_G(t) = y_o(t) + m(dW_A/dt)T_2 \tag{2.243}$$

Therefore, the signal $y_{om1}(t)$ at the output port of the multiplier, given by (2.220), becomes

$$y_{om1}(t) = [\mu_1|y_A(t)|^2 T_2]\frac{dW_A}{dt} + \mu_1|y_A(t)|^2 W_A + \frac{\sqrt{2}\mu_1|y_A(t)\,y_M^*(t)|}{m} \tag{2.244}$$

Note that due to the extra combiner in the $y_o(t)$ line and the extra splitter in the $y_A(t)$ line, the new value of μ_1 becomes

$$\mu_1 = \frac{GApm}{8}$$

Hence, (2.223) becomes

$$\frac{d^2 W_A}{dt^2} + \left[\frac{T_1 + T_2}{T_1 T_2} + \frac{\mu p_{Ai} T_2}{T_1 T_2}\right]\frac{dW_A}{dt} + \frac{1 + \mu p_{Ai}}{T_1 T_2} W_A = -\frac{\sqrt{2}\mu' v_{MAi}}{T_1 T_2} \tag{2.245}$$

where, now

$$\mu = \frac{GAgpm}{(\sqrt{2})4}$$

and, as before, $\mu' = \mu/pm$.

For $T_2 \gg T_1$, (2.245) reduces to

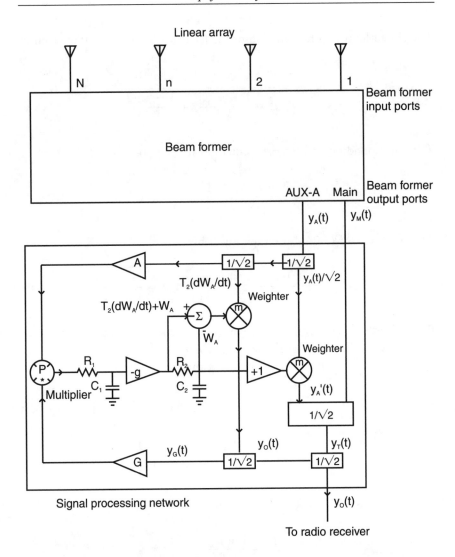

Figure 2.24 Nonoscillatory two-pole feedback loop.

$$\frac{d^2 W_A}{dt^2} + \left[\frac{1 + \mu p_{Ai}}{T_1}\right]\frac{dW_A}{dt} + \frac{1 + \mu p_{Ai}}{T_1 T_2} W_A = -\frac{(\sqrt{2})\mu' v_{MAi}}{T_1 T_2}$$

(2.246a)

Converting into complex frequency domain, (2.246) will give

$$W_A\left[s^2 + \frac{1 + \mu p_{Ai}}{T_1}s + \frac{1 + \mu p_{Ai}}{T_1 T_2}\right] = -\frac{(\sqrt{2})\mu' v_{MAi}}{T_1 T_2}$$

(2.246b)

The closed-loop transfer function of the feedback loop is therefore given by

$$H_{Ai} = -\frac{\sqrt{2}\mu'/(T_1 T_2)}{s^2 + \dfrac{1 + \mu p_{Ai}}{T_1}s + \dfrac{1 + \mu p_{Ai}}{T_1 T_2}} \tag{2.247}$$

The corresponding characteristic equation for the poles is

$$s^2 + \frac{1 + \mu p_{Ai}}{T_1}s + \frac{1 + \mu p_{Ai}}{T_1 T_2} = 0 \tag{2.248}$$

The two roots of (2.248) are then given by

$$s_{p1,p2} = -\frac{1 + \mu p_{Ai}}{2T_1} \pm \frac{1 + \mu p_{Ai}}{2T_1}\sqrt{1 - \frac{4T_1/T_2}{1 + \mu p_{Ai}}} \tag{2.249}$$

From (2.249), it is seen that for $T_2 > 4T_1$, the roots will always remain real irrespective of the value of the loop gain μp_{Ai}.

Equation (2.249) can be simplified to give

$$s_{p1,p2} = -\frac{1 + \mu p_{Ai}}{2T_1} \pm \frac{1 + \mu p_{Ai}}{2T_1}\left[1 - \frac{2T_1/T_2}{1 + \mu p_{Ai}}\right] \tag{2.250}$$

$$= -\frac{1 + \mu p_{Ai}}{2T_1} \pm \left[\frac{1 + \mu p_{Ai}}{2T_1} - \frac{1}{T_2}\right]$$

The two poles will therefore be given by

$$s_{p1} = -\frac{1 + \mu p_{Ai}}{T_1} + \frac{1}{T_2}$$

$$= -\frac{1 + \mu p_{Ai}}{T_1} \tag{2.251a}$$

$$s_{p2} = -\frac{1}{T_2} \tag{2.251b}$$

The root locus is shown in Figure 2.25. The pole s_{p1} is independent of loop gain and hence will remain stationary at $-(1/T_2)$. The pole s_{p2} will be at $-(1/T_1)$ for zero loop gain or if the feedback loop is disengaged. Then as

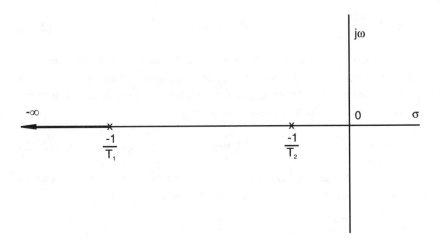

Figure 2.25 Root locus of a nonoscillatory two-pole feedback loop.

the loop gain increases, the pole s_{p2} will move toward $-$(infinity). However, s_{p1} is the dominant pole and hence the loop will remain unconditionally stable and the convergence time will essentially be independent of the loop gain.

2.13 Examples of Practical Beam Formers

In an adaptive array system, it is the beam former that receives the signals from the elements of the linear array and delivers them to the signal processing network with proper channeling. In this section a few practical beam formers are discussed in terms of their design, fabrication, and characteristics.

2.13.1 The Signals

Consider an N-element linear array, with $\lambda/2$ spacing between two adjacent array elements. There is a signal source located along an arbitrary direction $+\theta°$ off the array boresight. If the difference between the pathlengths traveled by the signal to reach array elements n and 1 is l_{n1}, then the phase difference between the signals at element n and element 1 is given by (1.2)

$$\psi_{n1} = \beta l_{n1} = (n - 1)\,\pi\sin\theta \quad n = 1, 2, \ldots, N \qquad (2.252)$$

and hence the phase difference between two adjacent array elements becomes (1.4b)

$$\psi = \pi \sin \theta \tag{2.253}$$

For element 1, $n = 1$. Hence, the phase at element 1 is zero and is regarded as the reference phase. Therefore, ψ_{n1} in (2.252) becomes the phase of the signal at element n relative to that at element 1, due to a signal source located at $+\theta°$ off the boresight and, hence,

$$\psi_{n1} = \psi_n = (n - 1) \pi \sin \theta \qquad n = 1, 2, \ldots, N \tag{2.254}$$

Note that if an initial phase ψ_1 is assumed to be associated with element 1, then the phase of the signal at element n will be given by

$$\psi_n = \psi_1 + (n - 1) \pi \sin \theta = \psi_1 + (n - 1)\psi \qquad n = 1, 2, \ldots, N \tag{2.255}$$

Again, from (2.253), it follows that if a signal source is located along the rth beam, at $+\theta_r°$ off the array boresight, then the phase difference between two adjacent array elements will be given by (1.6a)

$$\psi_r = \pi \sin \theta_r \tag{2.256}$$

and from (2.255), the general expression for the phase of the signal at element n due to a signal source along the rth beam will become

$$\psi_{nr} = \psi_1 + (n - 1) \pi \sin \theta_r = \psi_1 + (n - 1)\psi_r \qquad n = 1, 2, \ldots, N \tag{2.257}$$

Also, for a source located along $+\theta°$ off the boresight, the phase difference between the signals at two adjacent array elements, relative to the rth beam, will be given by (1.6b)

$$\psi_R = \psi - \psi_r \tag{2.258}$$

Again, when a signal generator is connected to the rth output port of an $N \times N$ beam former, the phase difference ψ_r between the signals at any two adjacent antenna ports will be (1.8)

$$\psi_r = \pi \sin \theta_r = (2\pi/N)r \tag{2.259}$$

Hence, θ_r, the direction of the rth beam relative to the array boresight will be given by

$$\theta_r = \sin^{-1}(\psi_r / \pi) = \sin^{-1}(2r/N) \qquad (2.260)$$

2.13.2   Typical 4 × 4 Beam Former

A 4 × 4 beam former has four input ports and four output ports as shown by the schematic diagram in Figure 2.26. The beam former is constructed by connecting four 180° hybrids and a 90° phase shifter as shown. For the hybrids, the Σ and Δ ports represent the sum ports and the difference ports, respectively. Thus, a signal traveling along the (−) diagonal path will suffer

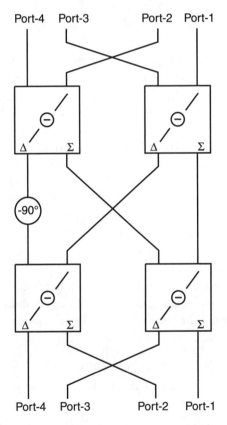

Antenna ports or input ports

Receiver ports or output ports

Figure 2.26 Typical 4 × 4 beam former.

a relative phase shift of 180° while there will be no relative phase shift along any other path.

2.13.2.1   Beam Locations for a 4 × 4 Beam Former

Let a four-element linear array be connected to the four input ports of the beam former. Then, for $N = 4$, from (2.259) and (2.260), the phase difference ψ_r between two adjacent input ports, due to a signal source located along the rth beam, at $+\theta_r$ off the boresight, will be given by

$$\psi_r = \pi r/2 \qquad (2.261)$$

and, hence,

$$\theta_r = \sin^{-1}(\psi_r/\pi) = \sin^{-1}(r/2) \qquad (2.262)$$

Now, for beam 0, $r = 0$ and hence $\psi_r = 0$ and $\theta_r = 0$. The beam is therefore located along the boresight with zero phase difference between any two adjacent input ports. Again, for beam-(+1), $r = +1$ and hence $\psi_r = +90°$ and $\theta_r = +30°$. Thus, the beam is located +30° off the boresight and the phase difference between two adjacent input ports is +90°. Similarly, for beam-(−1), $\psi_r = −90° = 270°$ and $\theta_r = −30°$ and, hence, the beam is located −30° off the boresight while the phase difference between two adjacent input ports is −90° or +270°. Finally, for beam-(+2), $r = 2$ and hence $\psi_r = +180°$ and $\theta_r = +90°$ and so the beam is located +90° off the boresight and the phase difference between two adjacent input ports is +180°. However, a phase shift of +180° or −180° brings a point to the same location, only along different routes. So, a phase shift of +180° associated with ψ_r could also be −180°. Consequently, according to (2.256), for $\psi_r = \pm180°$, the value of $\sin\theta_r$ could be either +1 or −1, indicating that beam 2 would simultaneously be located along +90° and −90°. This indicates that the array is operating at the endfire mode with two beams along the two ends of the horizontal axis of the array structure.

2.13.2.2   Phase Distribution at the Input Ports

If the initial phase at port 1 of the beam former is ψ_1, then for the rth beam, the phase values at the other input ports will be given by (2.257), where n will equal 1, 2, 3, and 4, corresponding to input ports 1, 2, 3, and 4, respectively.

Thus, for beam 0, $r = 0$ and hence by (2.259), $\psi_r = 0$. Then the signals at all the input ports will be in phase and equal to ψ_1, which is the phase at port 1. Again, for beam-(+1), $\psi_r = 90°$ and, hence, from (2.257), the phase at port 1 will still remain as ψ_1, but those at ports 2, 3, and 4 will be given by $\psi_1 + 90°$, $\psi_1 + 180°$, and $\psi_1 + 270°$, respectively. Similarly, for beam-(+2), $\psi_r = 180°$ and hence the phase values associated with the four input ports will be given by ψ_1, $\psi_1 + 180°$, $\psi_1 + 360°$, and $\psi_1 + 540°$, respectively. Finally, for beam-(−1), $\psi_r = 270°$ and so at the four input ports, the four phase values will be given by ψ_1, $\psi_1 + 270°$, $\psi_1 + 540°$, and $\psi_1 + 810°$, respectively. The phase value, associated with a given input port, for a given beam r, beam location θ_r, and the corresponding adjacent element phase difference ψ_r, are summarized in Table 2.14.

2.13.2.3 Input Port Signal Distribution

Let $v_r = V_r \angle 0$ represent the signal at the source located along the rth beam, where the initial phase of the signal at the source has been taken to be zero. Then assuming lossless propagation, the signal at the nth input port will become

$$v_{nr} = V_r \angle -\psi_1 - (n-1)\psi_r \qquad (2.263)$$

where, from (2.257), $[\psi_1 + (n-1)\psi_r]$ is the phase of the signal at the nth input port. Normalized with respect to V_r, (2.263) gives

$$v_{nr}/V_r = (v_{nr})_{norm} = 1 \angle -\psi_1 - (n-1)\psi_r \qquad (2.264)$$

Note that (2.264) also represents the transfer function or gain between a signal source located along the rth beam and the nth input port. Let the initial phase ψ_1 associated with the signal at element 1 be assumed to be zero, then (2.264) becomes

Table 2.14

r	θ_r	ψ_r	Input Port Phase Distribution			
			Port 1	Port 2	Port 3	Port 4
0	0°	0°	$-\psi_1$	$-\psi_1$	$-\psi_1$	$-\psi_1$
+1	+30°	90°	$-\psi_1$	$-\psi_1 - 90°$	$-\psi_1 - 180°$	$-\psi_1 - 270°$
+2	+90°	180°	$-\psi_1$	$-\psi_1 - 180°$	$-\psi_1 - 360°$	$-\psi_1 - 540°$
−1	−30°	−90° = 270°	$-\psi_1$	$-\psi_1 - 270°$	$-\psi_1 - 540°$	$-\psi_1 - 810°$

$$(v_{nr})_{norm} = 1 \big/ {-(n-1)\psi_r} \qquad (2.265)$$

Therefore, if we replace the phase values in Table 2.14 with their corresponding normalized signals, the resulting input port signal distribution will be as shown in Table 2.15.

The output ports, where the combined signals from the input ports for a given beam will appear, are also shown in Table 2.15. These output ports can be obtained by following the signal paths from the input end of the beam former to its output end.

2.13.2.4 Radiation Patterns of a 4 × 4 Beam Former

The signal at the rth output port of the beam former is the sum of the signals at the four input ports due to a signal source located along the rth beam. Hence, the signal at the rth output port represents the total transfer function or gain between the rth output port and the signal source along the rth beam. The normalized transfer function or gain can be obtained from (2.258) and (1.7a). For $N = 4$, the normalized gain will be given by

$$G_r(\theta) = \frac{\sin[4(\psi_R/2)]}{4\sin(\psi_R/2)} = \frac{\sin[4(\psi - \psi_r)/2]}{4\sin[(\psi - \psi_r)/2]} \qquad (2.266)$$

Plots of the output signals at the four output ports of the beam former, corresponding to the four beams, are shown in Figure 2.27 and represent the radiation patterns of the four beams.

It is noted that a given pattern has nulls along the peaks of all other patterns, indicating that the beams are orthogonal. Note that for clarity, the plots have been made using both positive and negative polarities.

Table 2.15

		Input Port Signal Distribution				
r	θ_r	Port 1	Port 2	Port 3	Port 4	Output Ports
0	0°	$1 \big/ 0°$	$1 \big/ 0°$	$1 \big/ 0°$	$1 \big/ 0°$	1
+1	+30°	$1 \big/ 0°$	$1 \big/ {-90°}$	$1 \big/ {-180°}$	$1 \big/ {-270°}$	4
+2	+90°	$1 \big/ 0°$	$1 \big/ {-180°}$	$1 \big/ {-360°}$	$1 \big/ {-540°}$	3
−1	−30°	$1 \big/ 0°$	$1 \big/ {-270°}$	$1 \big/ {-540°}$	$1 \big/ {-810°}$	2

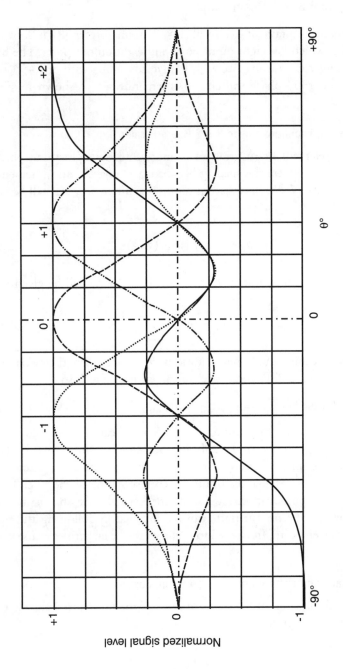

Figure 2.27 Radiation pattern of a 4 × 4 beam former between −90° and +90°. For clarity both negative and positive polarities have been used.

2.13.3 Typical 8 × 8 Beam Former

The schematic diagram of a typical 8 × 8 beam former [2] is shown in Figure 2.28, which has eight input ports and eight output ports. The beam former is constructed using twelve 180° hybrids, along with one 45°, three 90°, and one 135° phase shifters, all interconnected as shown in Figure 2.28.

2.13.3.1 Beam Locations for an 8 × 8 Beam Former

Let an eight-element linear array be connected to the eight input ports of the beam former where the spacing between two adjacent array elements is $\lambda/2$. Then, since $N = 8$, the adjacent port phase difference, ψ_r, will be given by (2.259)

$$\psi_r = (2\pi/N)r = (\pi/4)r \tag{2.267}$$

and from (2.260), the location of the +rth beam, +θ_r, will be given by

$$+\theta_r = \sin^{-1}(r/4) \tag{2.268}$$

The various beams r and their locations θ_r, along with their respective adjacent port phase differences ψ_r, are shown in Table 2.16.

2.13.3.2 Input Port Phase and Signal Distribution

The input port phase distributions of an 8 × 8 beam former, along with their respective beams, r, and the beam locations, θ_r, are shown in Table 2.17, where the initial phase ψ_1, at element 1, is assumed to be zero.

The input port signal distributions, corresponding to the given beams and the output ports carrying the combined signals, are shown in Table 2.18. As before, the output ports can be obtained by following the signal paths, for a given beam, from the input end of the beam former to its output end.

2.13.3.3 Radiation Patterns of an 8 × 8 Beam Former

Following (2.266), for an 8 × 8 beam former, the normalized signal at the rth output port is given by

$$G_r(\theta) = \frac{\sin[8(\psi - \psi_r)/2]}{8\sin[(\psi - \psi_r)/2]} \tag{2.269}$$

Antenna ports or input ports

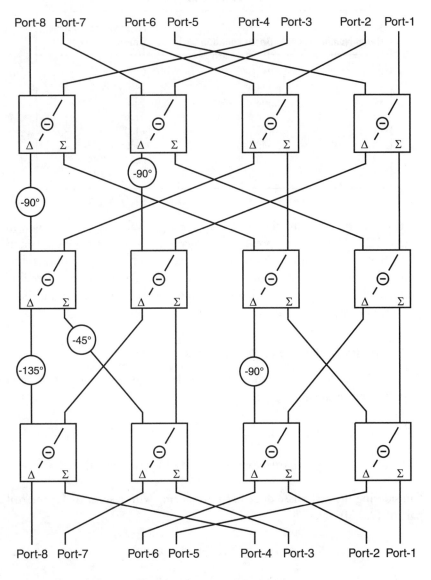

Receiver ports or output ports

Figure 2.28 Typical 8 × 8 beam former.

Table 2.16

Beam Number (r)	Beam Location (θ_r)	Adjacent Port Phase Difference (ψ_r)
0	0°	0°
+1	+14.5°	+45°
+2	+30°	+90°
+3	+48.6°	+135°
+4	+90°	+180°
−3	−48.6°	−135° = +225°
−2	−30°	−90° = +270°
−1	−14.5°	−45° = +315°

Plots of $G_r(\theta)$, as a function θ, for all the eight values of r are shown in Figure 2.29. These plots represent the radiation patterns of the eight beams of an 8 × 8 beam former. Note that the pattern of a given beam exhibits nulls along the peaks of all other beam patterns. The beams are therefore orthogonal. Here again, for clarity, the plots have been made using both positive and negative polarities.

2.13.4 Beam Formers with Real Signals

From (2.263), the signal at the nth array element due to a source, located along the rth beam, is

$$v_{nr} = V_r \angle -\psi_1 - n\psi_r \qquad (2.270)$$
$$= V_r[\cos(-\psi_1 - n\psi_r) + j\sin(-\psi_1 - n\psi_r)]$$

Considering the real phase only and normalizing with respect to V_r, the real part of (2.270) gives

$$\{v_{nr}\}_{\text{norm}} = \cos(-\psi_1 - n\psi_r) \qquad (2.271)$$

Similarly, the imaginary part of (2.270) gives

$$\{v_{nr}\}_{\text{norm}} = \sin(-\psi_1 - n\psi_r) \qquad (2.272)$$

Both (2.271) and (2.272) represent real values, which could either be positive or negative. Thus, when applied at the input ports of a 180° hybrid, they

Table 2.17

r	θ_r	Input Port Phase Distribution							
		Port 1	Port 2	Port 3	Port 4	Port 5	Port 6	Port 7	Port 8
0	0°	0°	0°	0°	0°	0°	0°	0°	0°
+1	+14.5°	0°	−45°	−90°	−135°	−180°	−225°	−270°	−315°
+2	+30°	0°	−90°	−180°	−270°	−360°	−450°	−540°	−630°
+3	+48.6°	0°	−135°	−270°	−405°	−540°	−675°	−810°	−945°
+4	+90°	0°	−180°	−360°	−540°	−720°	−900°	−1080°	−1260°
−3	−48.6°	0°	−225°	−450°	−675°	−900°	−1125°	−1350°	−1570°
−2	−30°	0°	−270°	−540°	−810°	−1080°	−1350°	−1620°	−1890°
−1	−14.4°	0°	−315°	−630°	−945°	−1260°	−1575°	−1890°	−2205°

Table 2.18

Beam Number (r)	Port 1	Port 2	Port 3	Port 4	Input Port Signal Distribution Port 5	Port 6	Port 7	Port 8	Output Ports
0	$1\angle 0°$	$1\angle 0°$	$1\angle 0°$	$1\angle 0°$	$1\angle 0°$	$1\angle 0°$	$1\angle 0°$	$1\angle 0°$	1
+1	$1\angle 0°$	$1\angle 45°$	$1\angle 90°$	$1\angle 135°$	$1\angle 180°$	$1\angle 225°$	$1\angle 270°$	$1\angle 315°$	8
+2	$1\angle 0°$	$1\angle 90°$	$1\angle 180°$	$1\angle 270°$	$1\angle 360°$	$1\angle 450°$	$1\angle 540°$	$1\angle 630°$	6
+3	$1\angle 0°$	$1\angle 135°$	$1\angle 270°$	$1\angle 405°$	$1\angle 540°$	$1\angle 675°$	$1\angle 810°$	$1\angle 945°$	7
+4	$1\angle 0°$	$1\angle 180°$	$1\angle 360°$	$1\angle 540°$	$1\angle 720°$	$1\angle 900°$	$1\angle 1080°$	$1\angle 1260°$	5
-3	$1\angle 0°$	$1\angle 225°$	$1\angle 450°$	$1\angle 675°$	$1\angle 900°$	$1\angle 1125°$	$1\angle 1350°$	$1\angle 1575°$	4
-2	$1\angle 0°$	$1\angle 270°$	$1\angle 540°$	$1\angle 810°$	$1\angle 1080°$	$1\angle 1350°$	$1\angle 1620°$	$1\angle 1890°$	2
-1	$1\angle 0°$	$1\angle 315°$	$1\angle 630°$	$1\angle 945°$	$1\angle 1260°$	$1\angle 1575°$	$1\angle 1890°$	$1\angle 2205°$	3

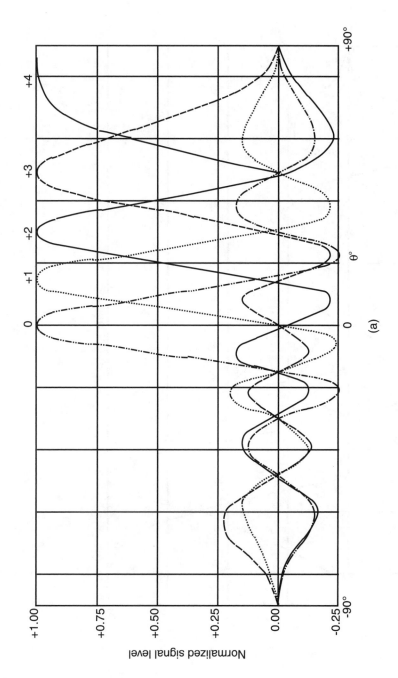

Figure 2.29 Radiation pattern of an 8 × 8 beam former between −90° and +90°. For clarity both positive (a) and negative (b) polarities have been used.

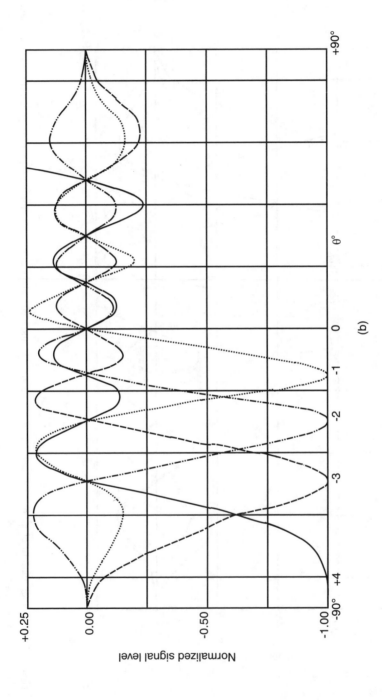

Figure 2.29 (continued).

will simply add or subtract along the way to an output port and will emerge as a real positive or negative quantity to be applied at an input port of the following hybrid. Therefore, if, instead of complex signals, real signals are used at the input ports of a beam former, then the extra phase shifters, used to construct the beam formers, could be omitted.

2.13.4.1 A 4 × 4 Beam Former with Real Input Signal Distribution

Using (2.271), the signal distribution at the input port of a 4×4 beam former presented in Table 2.15 could be modified as given in Table 2.19. A schematic diagram of the beam former, without the 90° phase shifter, is shown in Figure 2.30.

However, unlike the case of complex input signal distribution, discussed in Section 2.13.2.3, for a real input signal distribution, a choice of $\psi_1 = 0$ does not seem to produce a beam-forming effect. On the other hand, if the value of ψ_1 is chosen to be 45°, then a beam-forming effect will occur with real input signals. Thus, for $\psi_1 = 45°$, the real input port signal distribution and the corresponding output ports, where the combined signals will appear, are given by Table 2.20. The beam locations and the corresponding output ports are thus identical to those obtained with complex input port signal distribution given by Table 2.15.

It is worth noting that for beam 0, the phase difference ψ_r between two adjacent input ports is zero. Therefore, for beam 0, the initial phase ψ_1 for element 1 could also be set to zero, instead of 45° and so the input port signal distribution for beam 0 will be unity rather than 0.707. As a result, when real signals are used, the main beam, located along the array boresight, will be 3 dB higher than all the AUX beams.

2.13.4.2 An 8 × 8 Beam Former with Real Input Signal Distribution

The schematic diagram of an 8×8 beam former [3], to be used with real input port signals, is shown in Figure 2.31. Note that the beam former does

Table 2.19

			Input Port Signal Distribution			
r	θ_r	ψ_r	Port 1	Port 2	Port 3	Port 4
0	0°	0°	$\cos(-\psi_1)$	$\cos(-\psi_1)$	$\cos(-\psi_1)$	$\cos(-\psi_1)$
+1	+30°	+90°	$\cos(-\psi_1)$	$\cos(-\psi_1 - 90°)$	$\cos(-\psi_1 - 180°)$	$\cos(-\psi_1 - 270°)$
+2	+90°	+180°	$\cos(-\psi_1)$	$\cos(-\psi_1 - 180°)$	$\cos(-\psi_1 - 360°)$	$\cos(-\psi_1 - 540°)$
−1	−30°	−90° = 270°	$\cos(-\psi_1)$	$\cos(-\psi_1 - 270°)$	$\cos(-\psi_1 - 540°)$	$\cos(-\psi_1 - 810°)$

Antenna ports or input ports

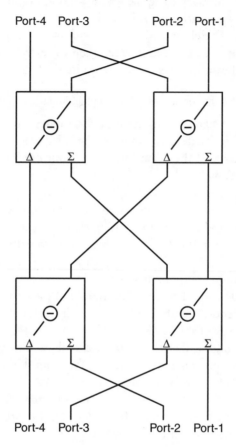

Receiver ports or output ports

Figure 2.30 A 4 × 4 beam former used with real input signals.

Table 2.20

Beam Number (r)	Beam Location (θ_r)	Input Port Signal Distribution				Output Port
		Port 1	Port 2	Port 3	Port 4	
0	0°	+0.707	+0.707	+0.707	+0.707	1
+1	+30°	+0.707	−0.707	−0.707	+0.707	4
+2	+90°	+0.707	−0.707	+0.707	−0.707	3
−1	−30°	+0.707	+0.707	−0.707	−0.707	2
0	0°	+1.000	+1.000	+1.000	+1.000	1

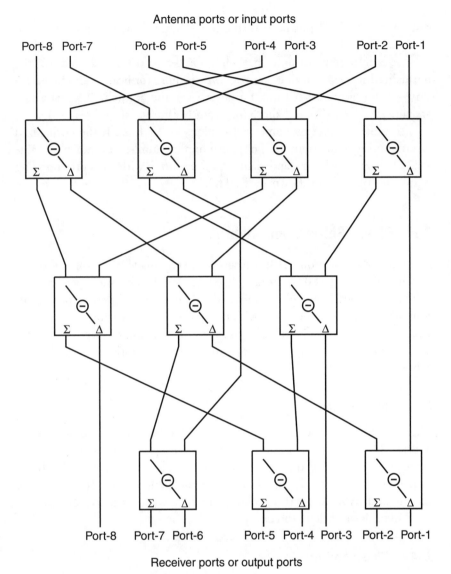

Figure 2.31 An 8 × 8 beam former used with real input signals.

not contain any phase shifters. Moreover, it contains only ten 180° hybrids compared to twelve for the beam former using complex input signals, as shown in Figure 2.28.

The complex normalized signals in Table 2.18 can now be replaced by real values obtained from (2.271) and (2.272). The resulting real signal distribution at the input ports of the 8 × 8 beam former of Figure 2.28 is presented in Table 2.21.

It is important to note that while for beams 0, +1, +2, and +3, the normalized real signal values are $\cos(-n\psi_r)$, those for beams −1 and −2 are $\sin(-n\psi_r)$. On the other hand, for beam-(−3), the normalized signal values are $\sin(-180° - n\psi_r)$ and those for beam-(+4) are $\cos(-45° - n\psi_r)$.

Evaluating the trigonometric functions in Table 2.21, the normalized input port signal distribution is obtained in pure numbers as shown in Table 2.22. The various beams and their locations, along with the corresponding output ports, are also shown in Table 2.22.

2.14 Phase Shifters and Hybrids

As discussed in Section 2.13, a beam former is made of phase shifters of various values and 180° hybrids. Therefore, for the optimum performance of a beam former, all the phase shifters should produce the desired phase shift over the desired frequency band with minimum deviation error. Likewise, the 180° hybrids should display minimum loss at the sum or Σ port and very high loss at the difference or Δ port. In this section, various types of phase shifters and 180° hybrids are discussed.

2.14.1 Introduction

All beam formers discussed in Section 2.13 were constructed with 180° hybrids. In addition, the beam formers with complex input signals also use extra phase shifters of different phase values. The performance of a beam former will therefore depend heavily on the performance of these phase shifters and hybrids. In this section, different types of phase shifters and 180° hybrids are briefly discussed.

2.14.2 Phase Shifters

A typical phase shifter is a four-port device where two in-phase signals entering at its two input ports will emerge at its two output ports with a constant differential phase shift over a desired frequency band. Two types of phase shifters are discussed here.

2.14.2.1 Schiffman Phase Shifters

Schiffman phase shifters [4–6] are relatively simple circuits which provide fairly constant differential phase shift between two output ports. A typical

Table 2.21

Beam Number (r)	Port 1	Port 2	Port 3	Input Port Real Signal Distribution Port 4	Port 5	Port 6	Port 7	Port 8
0	cos(0°)	cos(0°)	cos(0°)	cos(0°)	cos(0°)	cos(0°)	cos(0°)	cos(0°)
+1	cos(0°)	cos(−45°)	cos(−90°)	cos(−135°)	cos(−180°)	cos(−225°)	cos(−270°)	cos(−315°)
+2	cos(0°)	cos(−90°)	cos(−180°)	cos(−270°)	cos(−360°)	cos(−450°)	cos(−540°)	cos(−630°)
+3	cos(0°)	cos(−135°)	cos(−270°)	cos(−405°)	cos(−540°)	cos(−675°)	cos(−810°)	cos(−945°)
+4	cos(−45°)	cos(−225°)	cos(−405°)	cos(−585°)	cos(−765°)	cos(−945°)	cos(−1125°)	cos(−1305°)
−3	sin(−180°)	sin(−405°)	sin(−630°)	sin(−855°)	sin(−1080°)	sin(−1305°)	sin(−1530°)	sin(−1755°)
−2	sin(0°)	sin(−270°)	sin(−540°)	sin(−810°)	sin(−1080°)	sin(−1350°)	sin(−1620°)	sin(−1890°)
−1	sin(0°)	sin(−315°)	sin(−630°)	sin(−945°)	sin(−1260°)	sin(−1575°)	sin(−1890°)	sin(−2205°)

Table 2.22

r	θ_r	Input Port Signal Distribution								Output Port
		Port 1	Port 2	Port 3	Port 4	Port 5	Port 6	Port 7	Port 8	
0	0°	+1	+1	+1	+1	+1	+1	+1	+1	5
+1	+14.5°	+1	+0.707	0	-0.707	-1	-0.707	0	+0.707	2
+2	+30°	+1	0	-1	0	+1	0	-1	0	3
+3	+48.6°	+0.707	-0.707	0	+0.707	-1	+0.707	0	-0.707	1
+4	+90°	0	-0.707	+0.707	-0.707	+0.707	-0.707	+0.707	-0.707	4
-3	-48.6°	0	-0.707	+1	-0.707	0	+0.707	-1	+0.707	6
-2	-30°	0	+1	0	-1	0	+1	0	-1	8
-1	-14.5°	0	+0.707	+1	+0.707	0	-0.707	-1	-0.707	7

Schiffman phase shifter is shown in Figure 2.32. It contains a pair of parallel coupled transmission lines of length l_C, terminated by a matched load R_L and a section of 50-Ω uncoupled transmission line of length l_U, also terminated by a matched load R_L. If two coherent (in-phase) signals are applied at the input ports of l_C and l_U, they will appear at the output ports with a relatively constant phase difference over a fairly wide frequency band.

The phase shift ϕ_C caused by the coupled lines could be expressed as [4]

$$\cos\phi_C = \frac{(Z_{0e}/Z_{0o}) - \tan^2(\beta l_C)}{(Z_{0e}/Z_{0o}) + \tan^2(\beta l_C)} \tag{2.273}$$

where Z_{0e} and Z_{0o} are the even and odd mode characteristic impedances of the coupled lines. At the center frequency f_0 of the desired frequency band, the length l_C is quarter-wave or $\lambda_0/4$. Hence,

Inphase input signals

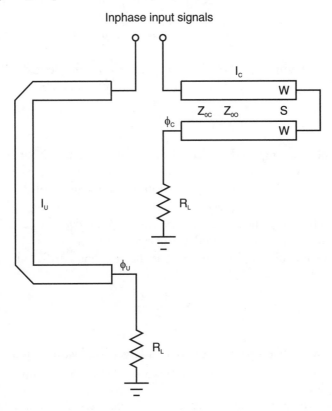

Figure 2.32 Typical Schiffman phase shifter.

$$\beta l_C = 2\pi\lambda_0/4\lambda = \pi f/2f_0 \qquad (2.274)$$

Therefore, (2.273) becomes

$$\cos\phi_C = \frac{(Z_{0e}/Z_{0o}) - \tan^2(\pi f/2f_0)}{(Z_{0e}/Z_{0o}) + \tan^2(\pi f/2f_0)} \qquad (2.275)$$

Again, the phase shift ϕ_U, caused by the uncoupled line l_U is

$$\phi_U = \beta l_U = 2\pi l_U/\lambda \qquad (2.276)$$

Therefore, the phase difference $\Delta\phi$, between the two output ports, is given by

$$\Delta\phi = \phi_U - \phi_C \qquad (2.277a)$$

$$\text{or} \quad \phi_U = \phi_C + \Delta\phi \qquad (2.277b)$$

From (2.275), at $f = f_0$, $\phi_C = 180°$. So, if at $f = f_0$, the desired differential phase of the given phase shifter is $\Delta\phi_0$, then from (2.277b),

$$\phi_{UO} = 180° + \Delta\phi_O \qquad (2.278)$$

Hence, the line length l_U is selected such that its phase shift at the center frequency is equal to 180° plus the desired phase shift $\Delta\phi_0$. For example, for a 90° phase shifter, the total phase shift due to the uncoupled line should be 270° at the center frequency. Then as the frequency changes, both ϕ_U and ϕ_C will change such that the difference $\Delta\phi$, as given by (2.277a), will remain close to $\Delta\phi_0$ or 90° with a small deviation error.

Now, for the coupled lines, if C is the coupling coefficient, then

$$\frac{Z_{0e}}{Z_{0o}} = \frac{1 + C}{1 - C} \qquad (2.279)$$

For a given midband phase shift $\Delta\phi_0$, a larger bandwidth requires tighter coupling. Likewise, for a given bandwidth, a larger $\Delta\phi_0$ also requires tighter coupling [6].

However, as the coupling coefficient C increases, the input return loss decreases [6]. In Figure 2.33, the coupling coefficient C has been plotted

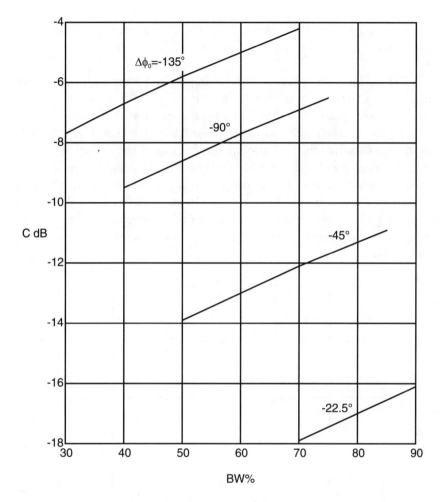

Figure 2.33 Bandwidth BW as a function of coupling coefficient C for different phase shifts $\Delta\phi_0$. The upper limits are the maximum obtainable bandwidths for phase error within ±5%.

against the fractional bandwidth BW for different values of midband phase shift $\Delta\phi_0$, for a maximum phase deviation of ±5%. In each case, the upper limit of the bandwidth represents the maximum obtainable bandwidth at that phase shift, for a phase error within ±5%.

Again, for phase shifters realized in microstrip configurations, the coupled lines are printed on a PC board where the line width w and coupling gap s are determined by the thickness h and relative dielectric constant ϵ_R of the PC board. The variation of the normalized gap s/h with the coupling coefficient C is shown in Figure 2.34, for ϵ_R = 6. For comparison, the midfrequency return loss RL as a function of C [6] is also shown in Figure 2.34.

Thus, for a 90° phase shifter, having a phase error within ±5%, the maximum achievable bandwidth is seen to be about 75% for a coupling coefficient of about −6.4 dB. This corresponds to a return loss of about −12 dB and a normalized gap s/h of about 0.055. Fabricated on a 0.0625-in.-thick PC board, this will correspond to a gap of about 0.0034 in.

On the other hand, for a 50% bandwidth, the coupling coefficient is about −8.6 dB, corresponding to a return loss of about −13.6 dB and a normalized gap of about 0.15. On a 0.0625-in.-thick PC board this will be equal to a gap of about 0.009 in. Note that for lower phase shifts, both the return loss and the gap dimension are adequate, even for maximum obtainable bandwidths.

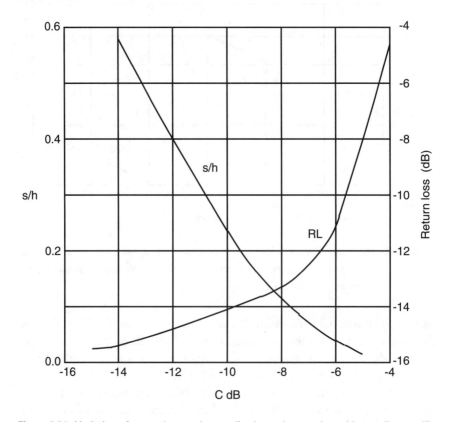

Figure 2.34 Variation of return loss and normalized coupler spacing with coupling coefficient.

Thus, phase shifters with smaller phase shifts can readily be constructed in Schiffman configurations. However, the fabrication of wideband large value phase shifters is restricted by poor return loss and small gap dimensions.

2.14.2.2 Halim Phase Shifters

If the pair of parallel coupled transmission lines in a Schiffman phase shifter is replaced by a pair of cross-coupled transmission lines, as shown in Figure 2.35, a Halim phase shifter [7] is obtained. Like a Schiffman phase shifter, a Halim phase shifter also contains a 50-Ω uncoupled transmission line l_U, terminated by a matched load R_L.

However, the cross-coupled section in a Halim phase shifter consists of a length of transmission line l_C with characteristic impedance Z_1 and is

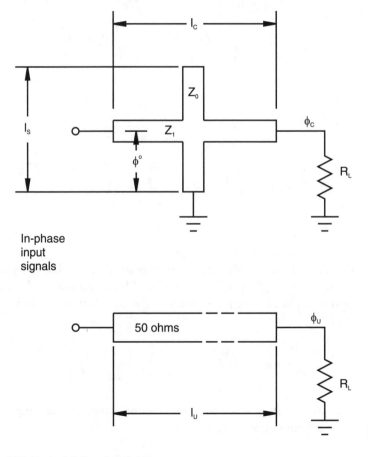

Figure 2.35 Typical Halim phase shifter.

half-wave long at the center frequency f_0. The line l_C is terminated by a matched load R_L. Perpendicular to l_C and located at its midpoint at $l_C/2$ is the second transmission line l_S, of characteristic impedance Z_0, which is quarter-wave long at f_0 and is shorted at one end.

The two lines l_C and l_S form the cross-coupled structure whose junction is $\theta°$ from the shorted end of l_S as shown in Figure 2.35.

If two in-phase signals are applied at the input ports of l_U and l_C, they will appear at their output ports with a fairly constant phase difference over a wide frequency band. At the junction of the cross coupled lines l_S and l_C, the phase α is given by [7, 8]

$$
\begin{aligned}
\tan \alpha &= \frac{Z_1^2}{Z_0^2} \frac{\pi Z_0}{8 R_L \sin^2 \theta} \left[\frac{f}{f_0} - \frac{f_0}{f} \right] \frac{x^2 + 1}{x^2 + 1/a^2} \sqrt{1 + \frac{x^2(a - 1/a)^2}{(x^2 + 1)^2}} \\
&= \frac{R_L}{Z_0} \frac{\pi}{8 \sin^2 \theta} \left[\frac{f}{f_0} - \frac{f_0}{f} \right] \frac{x^2 + 1}{x^2 a^2 + 1} \sqrt{1 + \frac{x^2(a^2 - 1)^2}{a^2(x^2 + 1)^2}} \quad (2.280) \\
&= k F(f)
\end{aligned}
$$

$$
\text{where} \quad k = \frac{R_L}{Z_0} \frac{\pi}{8 \sin^2 \theta} \quad (2.281)
$$

$$
\text{and} \quad F(f) = \left[\frac{f}{f_0} - \frac{f_0}{f} \right] \frac{x^2 + 1}{x^2 a^2 + 1} \sqrt{1 + \frac{x^2(a^2 - 1)^2}{a^2(x^2 + 1)^2}}
$$
$$
(2.282)
$$

with $a = R_L/Z_1$ and $x = \tan(\pi f/2f_0)$.

Thus, k is a function of the circuit parameters Z_0 and θ but independent of frequency while $F(f)$ is a function of the circuit parameter Z_1 and also a function of frequency. In particular, if $Z_1 = R_L$, then $a = 1$, and (2.280) reduces to

$$
\tan \alpha = k F(f) = \frac{R_L}{Z_0} \frac{\pi}{8 \sin^2 \theta} \left[\frac{f}{f_0} - \frac{f_0}{f} \right] \quad (2.283)
$$

Again, the phase shift ϕ_H due to a transmission line that is half wave at f_0, is given, as a function of frequency, by

$$
\phi_H = \beta \lambda_0/2 = \pi f/2 f_0 \quad (2.284)
$$

Hence, the total phase shift at the output of the cross-coupled transmission line l_C is

$$\phi_C = \phi_H - \alpha \qquad (2.285)$$

Again, the phase shift caused by the transmission line l_U is

$$\phi_U = \beta l_U = 2\pi l_U/\lambda \qquad (2.286)$$

Hence, the phase difference at the output ports of l_U and l_C is given by

$$\Delta\phi = \phi_U - \phi_C \qquad (2.287a)$$
$$\text{or} \qquad \phi_U = \Delta\phi + \phi_C \qquad (2.287b)$$

At $f = f_0$, $\alpha = 0$ and $\phi_H = 180°$. Then, from (2.285), $\phi_C = 180°$. So, if at $f = f_0$, the desired phase shift is designated $\Delta\phi_0$ then, from (2.287b),

$$\phi_{UO} = 180° + \Delta\phi_0 \qquad (2.288)$$

This equation is identical to (2.278). Hence, as before, at $f = f_0$, the length l_U is chosen to produce a total phase shift of $180°$ plus the desired phase shift $\Delta\phi_0$. Thus, for a $90°$ phase shifter, the phase shift by l_U should be $270°$ at f_0. Then, as the frequency changes, both ϕ_U and ϕ_C will change and $\Delta\phi$, given by (2.287a), will remain close to $90°$, with a small deviation error.

Since no coupled lines are involved, Halim phase shifters are relatively easy to fabricate. Moreover, optimization of the phase shifters could be achieved by adjusting the three circuit parameters, Z_1, Z_0, and θ.

Phase shifters with $90°$ phase shift have been built [7] to yield greater than 80% bandwidth with less than ±5% phase error, for a return loss of greater than 15 dB. Likewise, $135°$ phase shifters, having 70% bandwidth with less than 5% phase error and greater than 11 dB of return loss, have also been fabricated [7]. On the other hand, realization of $180°$ phase shifters having 80% bandwidth with return loss greater than 10 dB and a phase error under ±10% has also been reported [7]. Although an error of ±10% may not be acceptable for many applications, it nevertheless shows the close vicinity of the correct parameter values.

However, the accuracy of (2.280), diminishes at frequencies far away from the center frequency [8]. To obtain a fair assessment, the cross-coupled transmission line section has been analyzed using Touchstone software for different combinations of θ, Z_0, and Z_1 [7]. Tables of transmission loss S_{21}, return loss S_{11}, and the phase angle ϕ_C for three different values of θ are presented in Appendix J for various combinations of Z_0 and Z_1. They have been obtained by using Touchstone software at a center frequency of 2 GHz where the circuit is to be realized in microstrip configuration on a 0.062-in.-thick PC board of relative dielectric constant 2.2.

It is interesting to note that in order to realize a 180° phase shifter, $Z_1 = Z_0 = 25\Omega$ with $\theta = 30°$ appears to be a fairly good combination. On the other hand, to realize a 30° phase shifter, $\theta = 60°$ with $Z_0 = 76\Omega$ and $Z_1 = 32\Omega$, seems to be more appropriate. Thus, to obtain a phase shifter with large phase shift, a lower tap point on the shorted transmission line l_S is desirable. Then as the required phase shift decreases, the tap point moves upward. The rule of thumb therefore seems to be this: use a lower tap point with smaller line impedance for a larger phase shift, and a higher tap point with larger line impedance for a smaller phase shift.

A sample Touchstone program is shown in Figure J.2 for $Z_1 = Z_0 = 25\Omega$ and $\theta = 30°$. The related circuit parameters are shown in Figure J.1.

A Halim phase shifter could be optimized by noting that at $f = f_0$, the length of the uncoupled line is given by

$$l_U = \phi_{UO}/\beta_o \tag{2.289}$$

where $\beta_0 = 2\pi/\lambda_0$. Substitution of (2.289) into (2.286) gives

$$\phi_U = \beta l_U = \beta\phi_{UO}/\beta_O = \phi_{UO}\lambda_O/\lambda = \phi_{UO}(f/f_0) \tag{2.290}$$

Then, (2.287) yields

$$\phi_C = \phi_{UO}(f/f_0) - \Delta\phi \tag{2.291}$$

Equation (2.291) is an expression of the theoretical or ideal ϕ_C as a function of frequency, for a given phase shift of $\Delta\phi$. On the other hand, computations using Touchstone would give the actual values of ϕ_C corresponding to the given circuit configuration. Hence, during computation, the circuit parameters Z_1, Z_0, and θ could be varied so that the difference between the ideal and computed values of ϕ_C would be minimized.

Note, however, that although (2.283) could be used to realize phase shifters with larger phase values, for small value phase shifters, Z_0 could become too large for accurate practical realization.

2.14.2.3 Examples of Practical Phase Shifter Design

A few examples to design practical phase shifters have been given below.

To design a Halim phase shifter to a desired phase shift $\Delta\phi$, the θ, Z_1, and Z_0 combination has been selected from Appendix J, for which the phase ϕ_c of the cross-couples line is close to the ideal values having the maximum permitted phase deviation.

On the other hand, to design a Schiffman phase shifter to yield the desired phase shift $\Delta\phi$, the phase ϕ_c of the parallel coupled line has been calculated from the given bandwidth and the maximum tolerable phase deviation.

(a) Examples to Design Halim Phase Shifters

Example 2.21

Using the data from the Touchstone-generated table, Tables J.1 through J.6, design a 45° Halim phase shifter having a deviation error of less than 3% over a bandwidth of 50%.

Answer

For a 45° phase shifter, the phase ϕ_{uo} of the uncoupled line at the center frequency $f = f_0$, is $-180° - 45° = -225°$. Then the values of ϕ_u at different values of f/f_0 will be as shown in Table 2.23. Now, from Tables J.1 through J.6, a suitable combination of θ, Z_1, and Z_0 is selected such that the corresponding phase values ϕ_c, in conjunction with those of ϕ_u in Table 2.23, would produce a phase difference $\Delta\phi$ which is close to the desired value of 45°. Accordingly, Table J.4B is chosen for which the phase values ϕ_c, for different values of f/f_0 are shown in Table 2.23. The resulting phase difference $\Delta\phi$ for different values of f/f_0 and the corresponding deviation error are also shown in Table 2.23. We can see that the deviation error is less than 3% over a bandwidth greater than 60%.

Table 2.23

f/f_0	0.5	0.6	0.7	0.8	0.9	1.0	1.1	1.2	1.3	1.4	1.5
ϕ_u	−112.5	−125.0	−157.5	−180.0	−202.5	−225.0	−247.5	−270.0	−292.5	−315.0	−337.5
ϕ_c	−60.2	−87.8	−111.8	−134.4	−156.8	−179.4	−202.1	−224.7	−247.5	−271.8	−299.8
$\Delta\phi$	−52.3	−37.2	−43.7	−45.6	−45.7	−45.6	−45.4	−45.3	−45.0	−42.2	−37.7
Error	−7.3	+7.8	+1.3	−0.6	−0.7	−0.6	−0.4	−0.3	0.0	+2.8	+7.3

Example 2.22

Use Tables J.1 through J.6 to design a 90° Halim phase shifter that will have a deviation error less than 3% for a bandwidth over 60%.

Answer

For a 90° phase shifter, the phase ϕ_{u0} of the uncoupled line at the center frequency, $f = f_0$ is $-180° - 90° = -270°$. Then, for different frequencies, the values of ϕ_u will be as shown in Table 2.24. For the phase ϕ_c of the coupled line, the values in Table J.3A, corresponding to $\theta = 45°$ and $Z_1 = Z_0 = 32\Omega$ will be chosen. The values of ϕ_c for different frequencies are shown in Table 2.24. The resulting phase difference is given by $\Delta\phi$ in Table 2.24, which also shows the deviation error. We can see that the maximum deviation error is less than 2.5% for a bandwidth greater than 60%. The corresponding insertion loss and return loss are given in Table J.3A.

Example 2.23

Using the data given in Tables J.1 through J.7, design a 135° Halim phase shifter having a deviation of less than 3% over a bandwidth of 50%.

Answer

For a 135° phase shifter, the phase ϕ_{u0} of the uncoupled line at the center frequency $f = f_0$, is $-180° - 135° = -315°$. Hence, at different frequencies, the phase ϕ_u will be as given in Table 2.25. For the phase ϕ_c of the coupled line, Table J.1C, for $\theta = 30°$ and $Z_1 = Z_0 = 22.5\Omega$, will be chosen where ϕ_c at different frequencies are as shown in Table 2.25. The resulting phase difference $\Delta\phi$, along with the corresponding deviation error, are also shown in Table 2.25. Table 2.25 shows that the deviation error is less than 3% for a bandwidth greater than 50%. As before, the return loss and the insertion loss are given in Table J.1C.

Table 2.24

f/f_0	0.5	0.6	0.7	0.8	0.9	1.0	1.1	1.2	1.3	1.4	1.5
ϕ_u	−135	−162	−189	−216	−243	−270	−297	−324	−351	−378	−405
ϕ_c	−32.5	−68.0	−100	−128	−154	−180	−205	−232	−260	−292	−327.5
$\Delta\phi$	−102.5	−94	−89	−88	−89	−90	−92	−92	−91	−86	−77.5
Error	−10.5	−4.0	+1.0	+2.0	+1.0	0	−2.0	−2.0	−1.0	+4.0	+12.5

Table 2.25

f/f_0	0.5	0.6	0.7	0.8	0.9	1.0	1.1	1.2	1.3	1.4	1.5
ϕ_u	−157.5	−189.0	−220.5	−252.0	−283.5	−315.0	−346.5	−378.0	−409.5	−441.0	−472.5
ϕ_c	−3.0	−48.0	−88.0	−120.0	−150.0	−180.0	−210.0	−241.0	−279.0	−337.0	−385.0
$\Delta\phi$	−154.5	−141.0	−132.5	−132.0	−133.5	−135.0	−136.5	−137.0	−130.5	−104.0	−87.5
Error	−19.5	−6.0	+2.5	+3.0	+1.5	0.0	−1.5	−2.0	+4.5	+31.0	+47.5

Example 2.24

Use the data from Tables J.1 through J.6 to design a 180° Halim phase shifter that will produce a bandwidth in excess of 50%. Calculate the best maximum deviation error.

Answer

For a 180° phase shifter, the phase ϕ_{u0} of the uncoupled line at the center frequency $f = f_0$ is 180° + 180° = 360°. Then, at various frequencies, the phase ϕ_u will be as shown in Table 2.26. Now, from Tables J.1 through J.6, a suitable combination of θ, Z_1 and Z_0 will have to be chosen to yield the proper phase value ϕ_c of the uncoupled line. Since the phase shift is large, data for the lowest tap point or θ in Tables J.1 through J.6 will be considered. Accordingly, from Table J.1B, the phase ϕ_c, corresponding to the combination, $\theta = 30°$ and $Z_1 = Z_0 = 27.5\Omega$, as shown in Table 2.26, seems to be the most suitable one. The corresponding phase difference $\Delta\phi$ and the deviation error are also shown in Table 2.26. Note that the maximum deviation error is a little over 8% over a bandwidth of 100%. The return loss over the band seems to be acceptable. A lower tap point would probably improve the deviation error, which is left to the reader as an exercise.

Table 2.26

f/f_0	0.5	0.6	0.7	0.8	0.9	1.0	1.1	1.2	1.3	1.4	1.5
ϕ_u	−180	−216	−252	−288	−324	−360	−396	−432	−468	−506	−540
ϕ_c	−11	−49	−87	−121	−151	−180	−209	−238	−274	−332	−370
$\Delta\phi$	−169	−167	−165	−169	−173	−180	−187	−194	−174	−174	−170
Error	+11	+13	+15	+11	+7	0	−7	−14	+6	+6	+10

(b) Examples to Design Schiffman Phase Shifters

Example 2.25

Using Schiffman configuration, design a 90° phase shifter which will display a maximum phase deviation error of ±5% over a bandwidth of (i) 50% and (ii) 75%.

Answer

For a 90° phase shifter, the phase ϕ_u of the uncoupled line in a Schiffman phase shifter is −270° at f/f_0 = 1.0 and changes with frequency as shown in Table 2.27. Again, at f/f_0 = 1.0, the phase ϕ_c of the coupled line is −180° causing the desired phase shift $\Delta\phi_0$ to be −90°. For an ideal phase shifter, the desired phase shift $\Delta\phi$ remains −90° over the given frequency band so that the phase ϕ_c of the uncoupled line changes with frequency as shown in Table 2.27. However, for a practical phase shifter, ϕ_c experiences a certain amount of phase deviation as it changes with frequency, causing a corresponding phase deviation in $\Delta\phi$.

   In general, $\Delta\phi$ is −90° at the center frequency f_0, at a frequency f_L, which is lower than f_0 and at a frequency f_U which is higher than f_0. At frequencies beyond f_L and f_U, the value of $\Delta\phi$ deviates from −90°. Likewise, at the intermediate frequencies between f_0 and f_L and between f_0 and f_U, the value of $\Delta\phi$ also deviates from −90° (see Figure 3 in [4]).

   Note that at frequencies lower than f_L, the phase shift $\Delta\phi$ is smaller than the desired phase shift $\Delta\phi_0$ while at frequencies higher than f_U, the phase shift $\Delta\phi$ is larger than the desired phase shift $\Delta\phi_0$. On the other hand, at the intermediate frequencies between f_0 and f_L, the phase shift $\Delta\phi$ is larger than the desired phase shift $\Delta\phi_0$ while at the intermediate frequencies between f_0 and f_U, the phase shift $\Delta\phi$ is smaller than the desired phase shift $\Delta\phi_0$.

Table 2.27

f/f_0	0.5	0.6	0.7	0.8	0.9	1.0	1.1	1.2	1.3	1.4	1.5
ϕ_u	−135	−162	−189	−216	−243	−270	−297	−324	−351	−378	−405
$\Delta\phi$	−90	−90	−90	−90	−90	−90	−90	−90	−90	−90	−90
ϕ_c	−45	−72	−99	−126	−153	−180	−207	−234	−261	−280	−315

The frequencies beyond f_L and f_U, at which the maximum tolerable phase deviation occurs, are regarded as the band edge frequencies. The phase deviation at the intermediate frequencies between f_0 and f_L, and between f_0 and f_U could be smaller or equal to the maximum tolerable value but could never exceed it.

(i) Phase Shifter with 50% Bandwidth

For a 50% bandwidth, the band edge frequencies are $f/f_0 = 0.75$ and $f/f_0 = 1.25$ where the phase values ϕ_u of the uncoupled line are $-202.5°$ and $337.5°$, respectively. For a maximum phase deviation of $\pm5\%$, the phase shift $\Delta\phi$ at $f/f_0 = 0.75$ will be $-85.5°$ while that at $f/f_0 = 1.25$ will be $-94.5°$. Hence, the phase ϕ_c of the coupled line will be $-117°$ and $-243°$, respectively.

Now,

$$\cos(-117) = \cos(-243) = 0.4540$$

and

$$\tan^2(0.75 \times 90) = \tan^2(1.25 \times 90) = 5.8284$$

Hence, from (2.273)

$$Z_{0e}/Z_{0o} = 2.1887$$

and from (2.279)

$$C = 0.3728 = -8.6 \text{ dB}$$

Then, in a 50Ω system,

$$Z_{0e} = 74\Omega$$

and

$$Z_{0o} = 34\Omega$$

Printed on a PC board of dielectric constant $\epsilon_R = 6.0$ and thickness h, the normalized line width w/h and the coupling gap s/h of the coupled line will be about 1.165 and 0.150, respectively. For h = 0.0625″, the line width and the coupling gap will be about 0.073″ and 0.009″, respectively. The numbers quoted in Section 2.14.2.1 have been obtained from this example.

(ii) Phase Shifter with 75% Bandwidth

For a 75% bandwidth, the band edge frequencies are f/f_0 = 0.625 and f/f_0 = 1.375 where the phase values ϕ_u of the uncoupled line are −168.75° and −371.25°, respectively. For a maximum phase deviation of ±5%, the phase shift $\Delta\phi$ at f/f_0 = 0.625 will be −85.5° while that at f/f_0 = 1.375 will be −94.5°. Hence, the phase values ϕ_c of the coupled line will be −83.25° and −276.75°, respectively.

Now,

$$\cos(-83.25) = \cos(-276.75) = 0.1175$$

and

$$\tan^2(0.75 \times 90) = \tan^2(1.25 \times 90) = 2.2398$$

Hence, from (2.273),

$$Z_{0e}/Z_{0o} = 2.8364$$

and from (2.279)

$$C = 0.4787 = -6.4 \text{ dB}$$

Then, in a 50Ω system,

$$Z_{0e} = 84\Omega$$

and

$$Z_{0o} = 30\Omega$$

Printed on a PC board of dielectric constant ϵ_R = 6.0 and thickness h, the normalized line width w/h and the coupling gap s/h of the coupled line will be about 0.985 and 0.055, respectively. For h =

0.0625″, the line width and the coupling gap will be about 0.062″ and 0.003″, respectively. The numbers quoted in Section 2.14.2.1 have been obtained from this example.

However, there are other circuit configurations available to achieve parallel transmission line coupling. Some of them could be physically realized without much difficulty to produce much tighter coupling than that which could be obtained with edge coupling. It is therefore possible to have alternate designs of Schiffman phase shifters to produce larger phase shifts with smaller deviation errors.

2.14.2.4 Alternate Design of Schiffman Phase Shifters

It should be noted, however, that beside edge coupling, parallel coupling of transmission lines could also be realized by broad side coupling. Figure 2.32 shows the design of a Schiffman phase shifter using edge coupling. In broad side coupling the circuit is printed on both sides of a PC board where the amount of coupling depends on the common area between the top and bottom transmission lines. Using such a configuration, circuits with tighter couplings could conveniently be realized [5].

However, the fabrication of the circuit board, to realize broad side coupling, becomes somewhat more complicated. Moreover, since the coupling gap is now equal to the board thickness, the freedom of choice becomes restricted.

2.14.3 180° Hybrids

As discussed in Section 2.13, the beam formers are constructed with 180° hybrids. In this section design, construction and characteristics of a few different hybrids will be discussed.

2.14.3.1 Characteristics of 180° Hybrids

A 180° hybrid is a four-port device with two input ports and two output ports. Two coherent or inphase signals entering at the two input ports will emerge at one output port as the sum signal and at the other output port as the difference signal. An ideal hybrid will have perfect amplitude and phase balance with large adjacent port isolation over a wide frequency band.

2.14.3.2 Ring Hybrid or Rat Race

A conventional ring hybrid [3, 9, 10] is shown in Figure 2.36. It is a four-port device made of a closed ring of transmission line with characteristic impedance $Z_C = 70\Omega$ where the impedance Z_0 looking into each port is

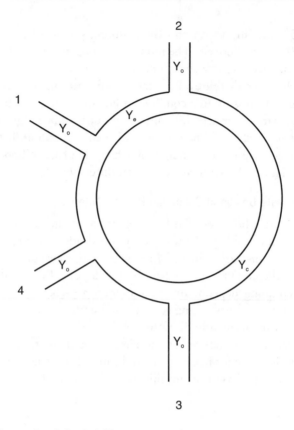

Figure 2.36 Conventional ring hybrid.

50Ω. The circumference of the ring is 1.5 wavelengths at the center frequency or $3\lambda_0/2$. Each of the lengths between ports 3–4, 4–1, and 1–2 is a quarter of a wavelength or $\lambda_0/4$ while that between ports 2–3 is three-quarters of a wavelength or $3\lambda_0/4$. Two in-phase signals entering ports 2 and 4 will come out at port 1 as the sum signal and at port 3 as the difference signal.

However, a conventional ring hybrid is inherently a narrowband device with a typical bandwidth of the order of 27% [9]. A conventional hybrid ring could be modified where the loop is divided into sections with unequal line impedance, which will increase the bandwidth [9].

A typical ring hybrid with unequal line impedance is shown in Figure 2.37, where Y_{1C} and Y_{2C} are quarter-wave transformers [8]. A bandwidth of 35% has been reported with a hybrid ring having impedance Z_1, Z_2, Z_3, Z_4, Z_{1C}, and Z_{2C} equal to 72, 62, 43, 33, 53, and 43Ω, respectively [3]. Tables presenting various sets of admittance values of a modified hybrid ring and the corresponding bandwidths are available [9].

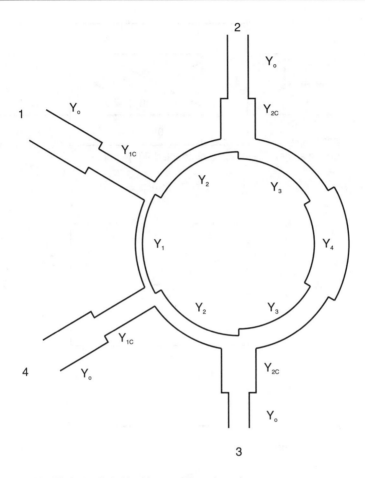

Figure 2.37 Modified ring hybrid with nonuniform impedances.

2.14.3.3   180° Hybrid from 90° Hybrid

From the above discussions, it becomes apparent that in applications where a bandwidth of more than 50% is required, a ring hybrid could not possibly be employed satisfactorily. In such cases, an alternate approach would be to cascade a wideband 90° phase shifter and a wideband 90° hybrid to build a wideband 180° hybrid. There are 90° hybrids available that are capable of producing quite large bandwidths. For example, Lange couplers [11] and slot couplers [12] are wideband 90° hybrids with bandwidths of the order of 2:1.

Figure 2.38 shows a 180° hybrid obtained by cascading a wideband phase shifter with a wideband 90° hybrid. However, from the figure it is

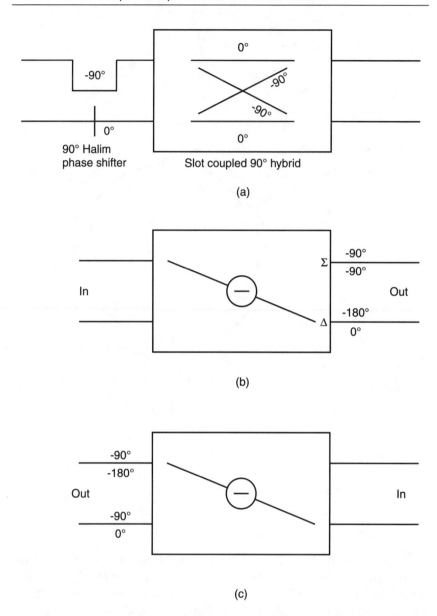

(a)

(b)

(c)

Figure 2.38 Nonreciprocal wideband 180° hybrid: (a) 90° phase shifter at the left side of 90° hybrid forming a nonreciprocal 180° hybrid; (b) signal flow from left to right in a nonreciprocal 180° hybrid; and (c) signal flow from right to left in a nonreciprocal 180° hybrid. [Note that the combination of the 90° phase shifter and the 90° hybrid in (a) has been represented by a single 180° hybrid in (b) and (c).]

seen that this hybrid is not a reciprocal device. Although it performs as a ring hybrid for signals propagating from left to right, it does not act as a 180° hybrid for signals propagating from right to left.

A reciprocal 180° hybrid is shown in Figure 2.39. Here two 90° phase shifters are placed, one on each side of a 90° hybrid. The performance of

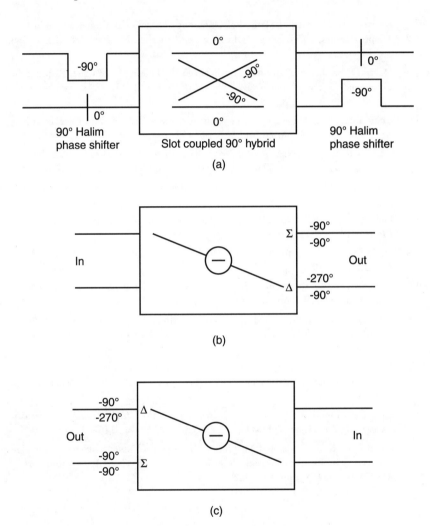

Figure 2.39   Reciprocal wideband 180° hybrid: (a) hybrid with a 90° phase shifter on each side forming a reciprocal 180° hybrid; (b) signal flow from left to right in a reciprocal 180° hybrid; and (c) signal flow from right to left in a reciprocal 180° hybrid.

this phase shifter is seen to be identical to that of a ring hybrid with a much larger bandwidth.

References

[1]   Klemes, M., "A Practical Method of Obtaining Constant Convergence Rate in LMS Adaptive Arrays," *IEEE Trans. on Antennas and Propagation,* Vol. AP-34, March 1986, pp. 440–446.

[2]   Davis, D. E. N., "A Transformation Between the Phasing Techniques Required for Linear and Circular Aerial Arrays," *Proc. IEE,* Vol. 112, No. 11, November 1965, pp. 2041–2045.

[3]   Abouzahra, M. D., "Design and Performance of a Wideband, Multilayer Feed Network," *Microwave J.,* November 1986, pp. 157–164.

[4]   Schiffman, B. M., "A New Class of Broad-Band Microwave 90-Degree Phase Shifters," *IRE Trans. on Microwave Theory and Techniques,* April 1958, pp. 232–237.

[5]   Kajfez, D., "Flatten the Response of Phase Shifters," *Microwaves,* December 1978, pp. 64–68.

[6]   Halim, M. A., "Easy Optimization of Phase Shifters in Microstrips," *Microwaves and RF,* July 1989, pp. 99–104.

[7]   Halim, M. A., "Resonator Circuits Expand Range of Phase Shifters," *Microwaves and RF,* December 1990, pp. 95–98.

[8]   *Reference Data for Radio Engineers,* 5th ed., Howard W. Sams & Co. Inc.: New York, A subsidiary of ITT Corp., pp. 13–15.

[9]   Kim, D., and Y. Yuki, "Broad-Band Design of Improved Hybrid-Ring 3-dB Directional Couplers," *IEEE Trans. on Microwave Theory and Techniques,* Vol. 30, No. 11, November 1982.

[10] Pon, C. Y., "Hybrid-Ring Directional Coupler for Arbitrary Power Divisions," *IEEE Trans. on Microwave Theory and Techniques,* Vol. MTT-9, November 1961, pp. 529–535.

[11] Lange, J., "Interdigitated Stripline Quadrature Hybrid," *IEEE Trans. on Microwave Theory and Techniques,* Vol. MTT-17, December 1969, pp. 1150–1151.

[12] Tanaka, T., K. Tsunoda, and M. Aikawa, "New Slot-Coupled Directional Couplers Between Double-Sided Substrate Microstrip Lines and Their Applications," *IEEE MTT-S Digest,* 1988, pp. 579–582.

3

Basic Matrix Expressions

Introduction

In general, an adaptive array system is a multisource, multioutput system as discussed in Section 2.8 and therefore could be represented by the block diagram shown in Figure 2.14.

Accordingly, being a multisource system, it involves a desired signal source located along the array boresight and a number of interference signal sources randomly distributed on both sides of the boresight (Figure 2.14).

Likewise, it is also a multioutput system where the system beam former contains an active main output port and a number of active AUX output ports.

To analyze such a multisource, multioutput system, the most suitable mathematical tools available involve matrix and vector algebra. This results in relatively complex mathematical expressions which may not be readily recognizable by some beginners.

In this chapter, some common complex matrix expressions contained in a standard book on adaptive arrays are explained in terms of related linear equations developed in Chapters 1 and 2. The process involves a few basic matrix relations found in many standard books of mathematical tables.

An in-depth treatment of the subject is given in [1], while [2] and [3] deal mainly with feedback loops.

A short list of the basic matrix relations used in this book is presented in Appendix I. These relations have been obtained from the book *Standard Mathematical Tables*, published by Chemical Rubber Company [4].

3.1 The Signals

An adaptive array system involves both desired and undesired signal sources distributed throughout space. Signals from all of these sources will travel through space to reach the linear array where they will be picked up by every array element and delivered to every input port of the beam former. The beam former will transport all signals to its output end and deliver them to all of its active output ports from where the signals will enter the signals processing network for the removal of undesired signals. The signals will then enter the radio receiver as a clean, desired signal, free of interference.

3.1.1 Distribution of Signal Sources and Signals

The system under consideration contains an $N \times N$ beam former whose input ports are connected to an N-element linear array (Figure 3.1).

There is a desired signal source located along the array boresight and transmitting the desired signal $f_d(t)$. There is also a number of interference signal sources which are randomly distributed on both sides of the boresight and transmitting interference signals $f_1(t)$, $f_2(t)$, $f_3(t)$, ...

It is assumed that none of the interference signal sources is located along any of the AUX beams of the beam former. It is also assumed that all signal sources, desired or undesired, are uncorrelated.

Each and every element of the N-element array will pick up signals from all the signal sources, desired and undesired, and will deliver them to the respective input port of the beam former. Thus, each and every input port of the beam former will carry signals from all signal sources.

3.1.2 Space Transfer Functions

As discussed in Section 2.1, let S_{nm} be the transfer function or gain between the mth interference signal source and the nth array element. Then (2.2)

$$S_{nm} = e^{-j\psi nm} \tag{3.1}$$

$$\text{where} \quad \psi_{nm} = (n - 1)\,\pi \sin\theta_m \tag{3.2}$$

is the phase of the signal received by array element n relative to element 1 from signal source m located along $+\theta_m$ off the boresight (2.1).

Let S_m represent the transfer functions between the mth signal source and all array elements; then from the mth column in Table 2.1,

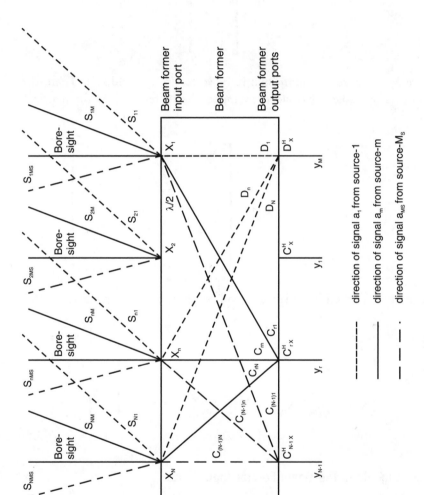

Figure 3.1 Multisource, multioutput system without signal processing network.

$$S_m = \begin{bmatrix} S_{1m} \\ S_{2m} \\ S_{3m} \\ \cdots \\ S_{nm} \\ \cdots \\ S_{Nm} \end{bmatrix} \tag{3.3}$$

where S_m is called a column matrix or vector (5, Appendix I). Then, the quantities in Table 2.1 could conveniently be represented by the matrix S given by

$$S = |S_1 \quad S_2 \quad \cdots \quad S_m| = \begin{bmatrix} S_{11} & S_{12} & \cdots & S_{1m} \\ S_{21} & S_{22} & \cdots & S_{2m} \\ S_{31} & S_{32} & \cdots & S_{3m} \\ \cdots & \cdots & \cdots & \cdots \\ S_{n1} & S_{n2} & \cdots & S_{nm} \\ \cdots & \cdots & \cdots & \cdots \\ S_{N1} & S_{N2} & \cdots & S_{Nm} \end{bmatrix} \tag{3.4}$$

Therefore, S^H, the Hermitian of S will be given by (3, Appendix I)

$$S^H = \begin{bmatrix} S_{11} & S_{21} & S_{31} & \cdots & S_{n1} & \cdots & S_{N1} \\ S_{12} & S_{22} & S_{32} & \cdots & S_{n2} & \cdots & S_{N2} \\ \cdots & \cdots & \cdots & \cdots & \cdots & \cdots & \cdots \\ S_{1m} & S_{2m} & S_{3m} & \cdots & S_{nm} & \cdots & S_{Nm} \end{bmatrix}^* = \begin{bmatrix} S_1^H \\ S_2^H \\ \cdots \\ S_m^H \end{bmatrix}$$

$$\tag{3.5}$$

3.1.3 Signals at the Beam Former Input Ports

As discussed in Section 2.1.2.2, if a_1, a_2, $\ldots$, a_m, are the amplitudes associated with signals transmitted by the interference signal sources 1, 2, $\ldots$, m, respectively, then the amplitudes of the signals at the nth array element due to the signals from the mth interference signal source will be given by (2.3a)

$$X_{nm} = S_{nm} a_m \tag{3.6a}$$

and that by the desired signal source will be given by (2.3b)

$$X_{nd} = S_{nd} a_d = a_d \tag{3.6b}$$

where S_{nd} is unity because the signals received by all array elements from the desired signal source are in phase ($\psi_{nd} = 0$). Therefore, the sum of the amplitudes of the signals from all signal sources, desired or undesired, at the nth input port of the beam former is (2.4)

$$X_n = X_{n1} + X_{n2} + \ldots + X_{nm} + \ldots + X_{nd} \tag{3.7}$$

$$= S_{n1} a_1 + S_{n2} a_2 + \ldots + S_{nm} a_m + \ldots + a_d$$

Considering the interference signals only, using Table 2.2, the amplitudes of the signals from all interference signal sources at all input ports of the beam former can be represented by

$$
\begin{bmatrix} X_1 \\ X_2 \\ \ldots \\ X_n \\ \ldots \\ X_N \end{bmatrix}
=
\begin{bmatrix}
S_{11} & S_{12} & S_{13} & S_{1m} & \ldots \\
S_{21} & S_{22} & S_{23} & S_{2m} & \ldots \\
\ldots & \ldots & \ldots & \ldots & \ldots \\
S_{n1} & S_{n2} & S_{n3} & S_{nm} & \ldots \\
\ldots & \ldots & \ldots & \ldots & \ldots \\
S_{N1} & S_{N2} & S_{N3} & S_{Nm} & \ldots
\end{bmatrix}
\begin{bmatrix} a_1 \\ a_2 \\ a_3 \\ \ldots \\ a_m \end{bmatrix}
\tag{3.8}
$$

$$\text{or} \qquad X = SA \tag{3.9}$$

$$
\text{where} \qquad X =
\begin{bmatrix} X_1 \\ X_2 \\ \ldots \\ X_n \\ \ldots \\ X_N \end{bmatrix}
\qquad
A =
\begin{bmatrix} a_1 \\ a_2 \\ a_3 \\ \ldots \\ a_m \end{bmatrix}
\tag{3.10}
$$

Thus, the column vector (5, Appendix I) X represents the amplitudes of all signals from all interference signal sources at all input ports of the beam former while column vector A represents the amplitudes of all signals transmitted by all interference signal sources.

Note, however, that in addition to the signals from all interference signal sources, all array elements will also receive signals from the desired signal source.

3.1.4 Beam Former Transfer Functions

As mentioned above, the beam former will transport all signals from its input end to its output end. However, there exists a finite transfer function between the input and output ports of the beam former that will modify the passing signals.

3.1.4.1 Main Port Transfer Function

Let the transfer functions or gains between the main output port and all input ports of the beam former (Table 2.3) be represented by the column vector

$$D = \begin{bmatrix} D_{M1} \\ D_{M2} \\ \cdots \\ D_{Mn} \\ \cdots \\ D_{MN} \end{bmatrix} \tag{3.11a}$$

Then, D^H, the Hermitian of D, will be given by

$$D^H = |D_{M1} \quad D_{M2} \quad \cdots \quad D_{Mn} \quad \cdots \quad D_{MN}|^* \tag{3.11b}$$

3.1.4.2 AUX Port Transfer Functions

Similarly, let the transfer functions or gains between all AUX output ports and all input ports of the beam former (Table 2.4) be given by the matrix

$$C = \begin{bmatrix} C_{11} & C_{21} & C_{31} & C_{r1} & C_{(N-1)1} \\ C_{12} & C_{22} & C_{32} & C_{r2} & C_{(N-1)2} \\ C_{13} & C_{23} & C_{33} & C_{r3} & C_{(N-1)3} \\ C_{1n} & C_{2n} & C_{3n} & C_{rn} & C_{(N-1)n} \\ C_{1N} & C_{2N} & C_{3N} & C_{rN} & C_{(N-1)N} \end{bmatrix} \tag{3.12a}$$

$$= |C_1 \quad C_2 \quad C_3 \quad C_r \quad C_{(N-1)}|$$

where C_r is the rth column matrix. Then C^H, the Hermitian of C, will be given by

$$C^H = \begin{bmatrix} C_{11} & C_{12} & C_{13} & C_{1n} & C_{1N} \\ C_{21} & C_{22} & C_{23} & C_{2n} & C_{2N} \\ C_{31} & C_{32} & C_{33} & C_{3n} & C_{3N} \\ C_{r1} & C_{r2} & C_{r3} & C_{rn} & C_{rN} \\ C_{(N-1)1} & C_{(N-1)2} & C_{(N-1)3} & C_{(N-1)n} & C_{(N-1)N} \end{bmatrix}^* \qquad (3.12b)$$

$$= \begin{bmatrix} C_1^H \\ C_2^H \\ C_3^H \\ C_r^H \\ C_{(N-1)}^H \end{bmatrix}$$

where C_r^H is the rth row matrix.

3.1.5 Signals at the Beam Former Output Ports

Due to the finite transfer function that exists between the input and output ports of a beam former, the signals appearing at its output ports will be those at the input ports modified by the transfer function.

3.1.5.1 Main Port Output Signals

From (2.7), the sum of the amplitudes of all signals appearing at the main output port of the beam former is

$$d_M = y_M = D_1^* X_1 + D_2^* X_2 + \ldots + D_n^* X_n + \ldots + D_N^* X_N \qquad (3.13)$$

$$= |D_1 \quad D_2 \quad D_n \quad D_N|^* \begin{bmatrix} X_1 \\ X_2 \\ X_n \\ X_N \end{bmatrix} = D^H X$$

Thus, $D^H X$ represents, in matrix notation, all signals from all input ports of the beam former, appearing at the main output port.

3.1.5.2 AUX Port Output Signals

From (2.11), the sum of the amplitudes of all signals appearing at the AUX-r output port of the beam former is

$$c_r = y_r = C_{r1}^* X_1 + C_{r2}^* X_2 + \ldots + C_{rn}^* X_n + \ldots + C_{rN}^* X_N \qquad (3.14)$$

$$= |C_{r1} \quad C_{r2} \quad C_{rn} \quad C_{rN}|^* \begin{bmatrix} X_1 \\ X_2 \\ X_n \\ X_N \end{bmatrix} = C_r^H X$$

When all AUX output ports are considered, (3.14) becomes

$$y_1 = C_1^H X$$
$$y_2 = C_2^H X \qquad (3.15)$$
$$y_r = C_r^H X$$
$$y_{(N-1)} = C_{(N-1)}^H X$$

which can now be expressed in matrix notation to give

$$\begin{bmatrix} y_1 \\ y_2 \\ y_r \\ y_{(N-1)} \end{bmatrix} = \begin{bmatrix} C_1^H X \\ C_2^H X \\ C_r^H X \\ C_{(N-1)}^H X \end{bmatrix} \qquad (3.16)$$

$$\text{or} \qquad Y_A = C^H X$$

where Y_A is the column vector representing all AUX output port signals and $C^H X$ represents, in matrix notation, all signals from all input ports of the beam former appearing at all AUX output ports.

Equations (3.13) and (3.16) are therefore the matrix representations of all signals at the main output port and all AUX output ports of the beam former and will now be delivered to the respective input ports of the signal processing network.

3.2 Signal Processing Network

The function of the signal processing network is to receive signals from all active output ports of the beam former, remove all undesired signals, and deliver the clean desired signal, free from interference, to the attached radio receiver.

3.2.1 Total Output Signal

As discussed in Chapter 2, signals from every AUX output port of the beam former will pass through a weighter (Figure 3.2). The output signals from all weighters are then combined with a combiner Σ to give the sum of the weighted signals.

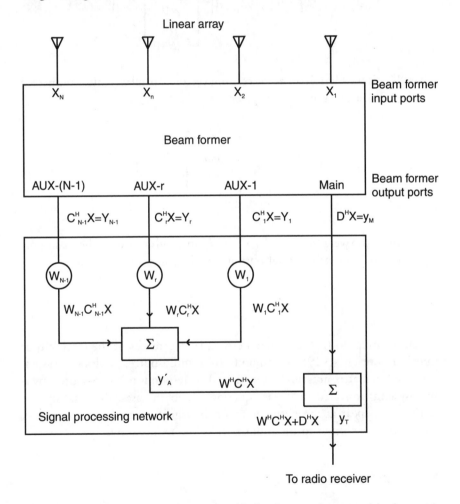

Figure 3.2 Multisource, multioutput system with signal processing network having manually adjustable weighters.

Let Y_A' represent the sum of the signals at the output ports of all the weighters as shown in Figure 3.2. Then

$$Y_A' = W_1^* C_1^H X + W_2^* C_2^H X + \ldots + W_r^* C_r^H X + \ldots$$
$$+ W_{(N-1)}^* C_{(N-1)}^H X \tag{3.17}$$

$$= |W_1 \quad W_2 \quad W_r \quad W_{(N-1)}|^* \begin{bmatrix} C_1^H \\ C_2^H \\ C_r^H \\ C_{(N-1)}^H \end{bmatrix} X = W^H C^H X$$

where W^H is the Hermitian of the column vector W of all weighters connected to all AUX output ports and is given by

$$W = \begin{bmatrix} W_1 \\ W_2 \\ W_r \\ W_{(N-1)} \end{bmatrix} \tag{3.18}$$

The sum of the weighted signals, Y_A', is again combined with the main port signal y_M to produce the total output signal y_T given by

$$y_T = Y_A' + y_M = W^H C^H X + D^H X \tag{3.19}$$

which is similar to (2.15a) where both equations represent the total output signal. However, (3.19) corresponds to a multisource, multiport system expressed in matrix notation while (2.15a) corresponds to a two-source, two-port system expressed in linear equations. In both cases, the total output signal will appear at the output port of the signal processing network and will then be delivered to the input port of the radio receiver.

3.2.2 Total Output Power

The total output power p_T can now be obtained from the total output signal y_T given by (3.19). Thus,

$$p_T = |y_T|^2 = |y_T y_T^*| = |y_T y_T^H|$$
$$= [W^H C^H X + D^H X][W^H C^H X + D^H X]^H$$
$$= [W^H C^H X + D^H X][X^H CW + X^H D] \qquad (3.20)$$
$$= [W^H C^H + D^H]XX^H[CW + D]$$
$$= [W^H C^H + D^H]R[CW + D]$$

$$\text{where} \quad R = XX^H \qquad (3.21)$$

and is known as the array covariance matrix. Since X represents the signals at all input ports of the beam former, from all transmitting signal sources, R will represent the powers of the signals from all signal sources at all input ports of the beam former. Equation (3.20) could therefore be written as

$$p_T = W^H C^H RCW + W^H C^H RD + D^H RCW + D^H RD \qquad (3.22)$$

Note that (3.22) is similar to (2.32), where each has four similar terms in the right-hand side. Although the terms in (3.22) are in matrix notations and those in (2.32) are single valued, each term has identical physical significance. Thus,

(a) $W^H C^H RCW$ in (3.22) corresponds to

$$\frac{m^2 W_A^2}{2}|G_A(\theta_i)|^2 p_i$$

in (2.32) and hence represents the weighted interference signal powers from all AUX output ports.

(b) $W^H C^H RD$ in (3.22) corresponds to

$$\frac{2m W_A}{\sqrt{2}}|G_M(\theta_i) G_A(\theta_i)| p_i$$

in (2.32) and represents the weighted cross-gain interference signal powers between the main output port and all active AUX output ports.

(c) $D^H RCW$ in (3.22) corresponds to $|G_M(\theta_i)|^2 p_i$ in (2.32) and represents the total interference signal power at the main output port.

(d) $D^H RD$ in (3.22) corresponds to $|G_M(0)|^2 p_d$ in (2.32) and represents the total desired signal power at the main output port.

3.2.3 Cancellation of Interference Signals

To remove the interference signals, the signal processing network will modify, with the help of a weighter, the amplitude and phase of each AUX port signal from the beam former so that when they all are combined with the main port signal, the interference signals present in the main port signal will be cancelled, leaving the desired signal only.

3.2.3.1 Adjusting the Weighters for Minimum Output Power

As discussed in Section 1.5.4, one way to remove the interference signal is to adjust the weighter for the minimum value of the total output power p_T. When p_T reaches its minimum value, that means all interference signals have been removed and weighters have reached their steady-state values. Thus, differentiating p_T with respect to W, (3.22) gives

$$\frac{\partial p_T}{\partial W} = 2C^H RCW + 2C^H RD \qquad (3.23)$$

Note that, looking at the expression of the total power given by (3.22), the appearance of the factor 2 in the right-hand side of (3.23) does not seem to be quite obvious. However, a look at the comparison between (3.22) and (2.32), using terms a and b presented under (3.22), would reveal that the term $W^H C^H RCW$ in (3.22) contains W_A^2 and the term $W^H C^H RD$ contains $2W_A$ where both will produce a factor of 2 on differentiation.

When p_T becomes minimum, $\partial p_T / \partial W$ becomes zero and (3.23) gives

$$C^H RCW + C^H RD = 0$$

$$\text{or} \quad W = -[C^H RC]^{-1} C^H RD \qquad (3.24)$$

Equation (3.24) gives the value of the weighters for which p_T will become minimum and the interference signals will be removed and therefore corresponds to the steady-state values of the weighters. Thus, if in Figure 3.2, all weighters could be set to the values given by (3.24), then all interference signals from the total output signal y_T can be removed. Note that (3.24) is similar to (1.24).

3.2.3.2 Adjusting the Weighters for Steady-State Values

Following the discussions in Section 1.5.4, the weighter derivative of p_T in (3.23) could also be regarded as the time derivative of the weighter. Hence, (3.23) will give

$$\frac{dW}{dt} = 2[C^H RCW + C^H RD]$$ (3.25)

At the steady-state values of the weighters, (3.25) will become zero and hence will give

$$W_{SS} = -[C^H RC]^{-1} C^H RD$$ (3.26)

3.2.3.3 Alternate Expression for the Time Derivative of the Weighters

The right-hand side of (3.25) can be expressed as

$$C^H RCW + C^H RD = C^H X X^H CW + C^H X X^H D$$
$$= C^H X | W^H C^H X |^H + C^H X | D^H X^H \quad (3.27)$$
$$= C^H X | W^H C^H X + D^H X |^H$$

Substitution of (3.19) into (3.27) gives

$$C^H RCW + C^H RD = C^H X y_T^H$$ (3.28)
$$= C^H X y_T^*$$

Again, substitution of the value of $C^H X$ from (3.16) into (3.28) gives

$$C^H RCW + C^H RD = Y_A y_T^*$$ (3.29)

Therefore, (3.25) becomes

$$\frac{dW}{dt} = 2 Y_A y_T^*$$ (3.30)

which is identical to (1.31) and, therefore, following Section 1.5.4, could be translated into a practical circuit as shown in Figure 3.3.

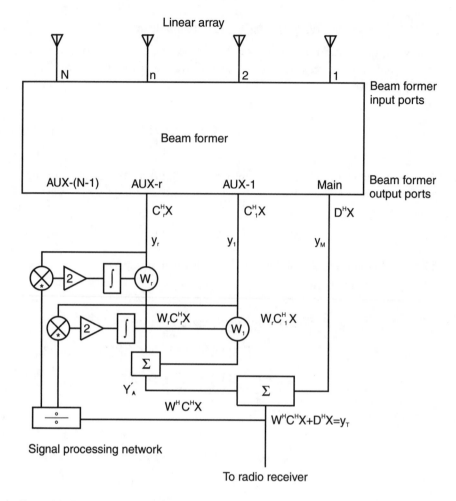

Figure 3.3 Basic multisource, multiport adaptive array with matrix notations.

3.3 Modal Representation of Matrix Expressions

To analyze a mathematical expression presented in matrix form is a difficult task. It becomes much easier if these matrix expressions can be represented in linear forms. Accordingly, the matrix expression under consideration is broken up into a number of linear expressions called *modal expressions*. Each modal expression can then be conveniently analyzed.

3.3.1 Linear Expressions for Matrix Equations

In (3.22), the total output power p_T has been expressed in terms of matrix notations and involves signals from all transmitting signal sources appearing at all active output ports of the beam former.

However, to obtain the final steady-state output power, (3.22) will have to be solved after replacing all weighters by their steady-state values given by (3.26).

Moreover, to study the feedback loops, the expression for the time derivative of the weighter, given by (3.25) will have to be solved. Evidently, it will be extremely difficult to solve these equations while they are still expressed in matrix notations.

Therefore, in order to use these equations for any effective practical applications, all terms expressed in matrix notations will have to be broken into a set of linear equations. Each of these linear equations could then be treated as discussed in Chapter 2.

The process of breaking up a matrix equation into a set of linear equations is called *modal decomposition,* where each linear equation is a modal representation of the matrix expression.

3.3.2 Modal Decomposition of the Matrix Expressions

As an example, we now decompose a few matrix expressions into their respective modal representations.

3.3.2.1 Decomposition of $C^H RC$

From (3.2) we can see that C^H is an $(N-1) \times N$ matrix having $(N-1)$ rows and N columns while C is an $N \times (N-1)$ matrix with N rows and $(N-1)$ columns.

Again, from (3.10), X is an N-element column matrix and, hence, from (3.21), R is an $N \times N$ square matrix (6, Appendix I). Then, $C^H R$ is an $(N-1) \times N$ matrix and $C^H RC$ is an $(N-1) \times (N-1)$ matrix. Put

$$C^H RC = M \qquad (3.31)$$

Then, since M is an $(N-1) \times (N-1)$ square matrix, it will satisfy the relation (9, Appendix I)

$$|M - \lambda I| = 0 \qquad (3.32)$$

where I is a unit matrix (8, Appendix I). Equation (3.20) is known as the *characteristic equation* of matrix M which has $(N-1)$ characteristic roots or eigenvalues given by (10, Appendix I)

$$\lambda = \lambda_1, \lambda_2, \ldots, \lambda_n, \ldots, \lambda_{(N-1)} \qquad (3.33)$$

These eigenvalues are obtained from the determinant of (3.32). The nth eigenvalue λ_n represents the nth mode of the matrix M.

Again, since λ_n is the nth eigenvalue of M, they will satisfy the relation (11, Appendix I)

$$Me_n = \lambda_n e_n \tag{3.34}$$

where e_n is a column matrix or vector, given by

$$e_n = \begin{bmatrix} e_{1n} \\ e_m \\ e_{(N-1)n} \end{bmatrix} \tag{3.35}$$

and represents the nth eigenvector associated with the nth eigenvalue λ_n of the matrix M. Therefore, the full matrix E of elements e could be expressed by a row matrix of the column matrix or the eigen vector e_n and will be given by (12, Appendix I)

$$E = |e_1 \quad e_2 \quad \cdots \quad e_n \quad \cdots \quad e_{(N-1)}| \tag{3.36}$$

Thus, E is an $(N-1) \times (N-1)$ square matrix. Hence, if Λ is the diagonal matrix of the eigenvalues λ, then (13, Appendix I)

$$M|e_1 \quad e_2 \cdots e_n \cdots e_{(N-1)}|$$

$$= |e_1 \quad e_2 \cdots e_n \cdots e_{(N-1)}| \begin{bmatrix} \lambda_1 & 0 & 0 & 0 \\ 0 & \lambda_2 & 0 & 0 \\ \cdots & \cdots & \cdots & \cdots \\ 0 & 0 & \lambda_n & 0 \\ \cdots & \cdots & \cdots & \cdots \\ 0 & 0 & 0 & \lambda_{(N-1)} \end{bmatrix}$$

$$= |e_1\lambda_1 \quad e_2\lambda_2 \quad \cdots \quad e_n\lambda_n \quad \cdots \quad e_{(N-1)}\lambda_{(N-1)}| \tag{3.37a}$$

$$\text{or} \quad ME = E\Lambda \tag{3.37b}$$

Again, the matrix M is symmetric and hence (14, Appendix I),

$$M = E\Lambda E^H \tag{3.38}$$

Also, since M is symmetric (15, Appendix I),

$$E = E^{-1}$$

$$\text{or} \quad EE^H = E^H E = I \tag{3.39}$$

where I is the unit matrix (8, Appendix I). Hence, (3.38) gives

$$M = E\Lambda E^H = E\Lambda E^{-1} \tag{3.40}$$

which represents the mode matrix of M.

3.3.2.2 Decomposition of $C^H RD$

As mentioned above, C^H is an $(N - 1) \times N$ matrix while R is an $N \times N$ square matrix. Hence, $C^H R$ is an $(N - 1) \times N$ matrix (6, Appendix I). Again, from (3.11a), D is an N-element column matrix or an $N \times 1$ matrix and, hence, $C^H RD$ is an $(N - 1)$-element column matrix. Therefore, if B represents the modes of $C^H RD$ then EB will represent its mode matrix. Therefore, since E is an $(N - 1) \times (N - 1)$ square matrix, B must be an $(N - 1)$-element column matrix. Put

$$C^H RD = Q \tag{3.41}$$

Then,

$$Q = EB = |e_1 \quad \cdots \quad e_n \quad \cdots \quad e_{(N-1)}| \begin{bmatrix} b_1 \\ \cdots \\ b_n \\ \cdots \\ b_{(N-1)} \end{bmatrix} \tag{3.42}$$

Thus, (3.42) represents the mode matrix of Q or $C^H RD$ where b_n represents the nth mode of Q.

3.3.2.3 Decomposition of $D^H RC$

From (3.11b), D^H is an N-element row matrix or a $1 \times N$ matrix. Then, since R is an $N \times N$ square matrix, $D^H R$ is a $1 \times N$ row matrix. Again, C is an $(N - 1) \times N$ matrix and hence $D^H RC$ is a $1 \times (N - 1)$ row matrix or an $(N - 1)$-element row matrix.

Let $D^H RC$ be represented by F. Then F is a $1 \times (N-1)$ row matrix and hence, EF will give the corresponding mode matrix. However, E is an $N \times N$ square matrix and hence the product EF could not be obtained (6, Appendix I). Thus, F by itself will also represent the mode matrix of $D^H RC$.

Note that $D^H RC$ corresponds to the signal powers at the main output port from all interference signal sources. However, the maximum number of elements in matrix $D^H RC$ is $(N-1)$ and hence, the maximum number of interference signal sources must also be $(N-1)$. Therefore, the maximum number of interference signal sources is equal to the maximum number of active AUX output ports.

Hence, the maximum number of interference signals that could effectively be treated by an adaptive array system is equal to the maximum number of active feedback loops present in the system.

3.3.2.4 Decomposition of $D^H RD$

As mentioned above, $D^H R$ is a $1 \times N$ row matrix. Then, since D is an N-element column matrix or $N \times 1$ matrix, $D^H RD$ is a single-valued quantity and represents the total desired signal power at the main output port.

3.3.2.5 Decomposition of the Weighter W

From (3.18), W is an $(N-1)$-element column matrix. Then, if L represents the modes of W, then EL will give the mode matrix. Therefore, since E is an $(N-1) \times (N-1)$ square matrix, L must be an $(N-1)$-element column matrix. Thus,

$$W = EL = \begin{vmatrix} e_1 & \cdots & e_n & \cdots & e_{(N-1)} \end{vmatrix} \begin{bmatrix} l_1 \\ \cdots \\ l_n \\ \cdots \\ l_{(N-1)} \end{bmatrix} \tag{3.43}$$

where, as before, (3.43) represents the mode matrix of W where l_n represents the nth mode of W.

3.3.3 Steady-State Values of the Weighters in Modal Notation

Figure 3.4 shows the rth feedback loop of a multisource, multiport system. As discussed in Chapter 2, the signal y_r at the rth AUX output port is divided into two equal parts $[y_r/\sqrt{2}]$ where one part is delivered to one of the two input ports of a multiplier through a gain block with gain A and

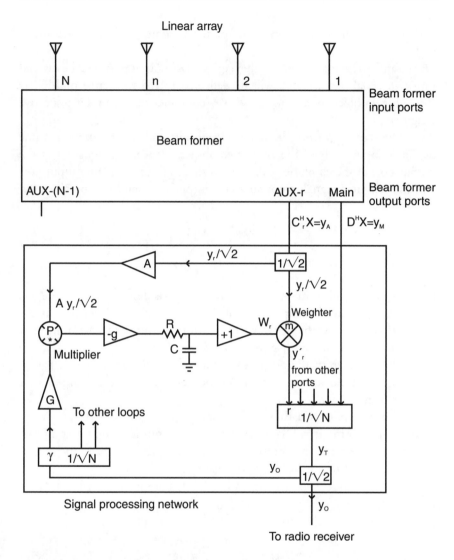

Figure 3.4 Practical multisource, multiport adaptive system. For clarity only the rth AUX port is shown.

the other part is applied at the input port of the weighter W_r. The signal y_r' from the output port of the weighter is applied to the rth input port of an N-way combiner where the signal y_r' is given by

$$y_r' = \frac{mW_r}{\sqrt{2}} \tag{3.44}$$

with m being the conversion factor of the weighter with a dimension of volt^{-1}.

Similarly, signals from all other active AUX output ports will be divided into two equal parts, which will be delivered to the respective multipliers through gain blocks A and to the N-way combiner through their respective weighters.

The signal y_M from the main output port will also be applied at one of the N input ports of the N-way combiner. Hence, the signal y_T at the output port of the combiner will contain signals from the main output port and the weighted signals from all AUX output ports and will be given by

$$y_T = \frac{y_r' + y_M}{\sqrt{N}} = \frac{mW_r}{\sqrt{(2N)}} + \frac{y_M}{\sqrt{N}} \qquad (3.45)$$

Signal y_T will again be divided into two equal parts designated y_0 where one part will appear at the output port of the signal processing network as the final output signal to be delivered to a radio receiver while the other part will be applied at the input port of an N-way divider.

The signal from the rth output port of the N-way divider is passed through a gain block with gain G and delivered to the second input port of the rth multiplier.

The multiplier will thus receive signals $[Ay_r/\sqrt{2}]$ and $[Gy_0/\sqrt{N}]$ at its two input ports. It will convert the signal y_0 into its complex conjugate y_T^* and will perform the multiplication process where the product signal will be delivered to a gain block with gain $-g$.

At the output port of the gain block $-g$, the signal y_{mo} will therefore be given by

$$y_{mo} = -g \frac{[Apy_r]}{\sqrt{2}} \frac{[Gy_0^*]}{\sqrt{N}} \qquad (3.46)$$

$$= -\frac{GApg}{2\sqrt{N}}[y_r y_T^*]$$

where p is the conversion factor of the multiplier with the dimension of volt^{-1}.

Therefore, if the signals from all AUX output ports are considered, y_r can be replaced by Y_A (3.16); hence, (3.16) will become

$$y_{om} = -\frac{GAgp}{2\sqrt{N}}[Y_A y_T^*] \qquad (3.47)$$

Substituting (3.29) into (3.47) and using (3.45) (the gains associated with signals from an AUX output port and the main output ports are $[m/\sqrt{(2N)}]$ and $[1/\sqrt{N}]$, respectively), (3.47) will give

$$
\begin{aligned}
y_{om} &= -\frac{GAgp}{2\sqrt{N}}\left[\frac{mC^HRCW}{\sqrt{(2N)}} + \frac{C^HRD}{\sqrt{N}}\right] \\
&= -\frac{GAgpm}{(2\sqrt{2})N}\left[C^HRCW + \frac{\sqrt{2}}{m}C^HRD\right] \qquad (3.48) \\
&= -\mu\left[C^HRCW + \frac{\sqrt{2}}{m}C^HRD\right] \\
&= -\mu\left[MW + \frac{\sqrt{2}}{m}Q\right]
\end{aligned}
$$

$$\text{where} \qquad \mu = \frac{GAgpm}{(2\sqrt{2})N} \qquad (3.49)$$

which is the total gain factor associated with the feedback circuit controlling the weighter and has the dimension of power^{-1}, while m and p are the conversion factors of the weighter and the multiplier, respectively, with the dimension of volt^{-1}. M and Q are given by (3.31) and (3.41), respectively.

Again, at the input port of the RC integrating circuit, the signal y_{ii} is given by

$$y_{ii} = RC\frac{dW}{dt} + W \qquad (3.50)$$

where W is given by (3.18) and represents all weighters involved. Put $RC = T$. Then, combining (3.46) and (3.48) gives

$$\frac{dW}{dt} = -\frac{(I + \mu M)W}{T} - \frac{\mu\sqrt{2}}{mT}Q \qquad (3.51)$$

where I is a unit matrix.

At steady state, $dW/dt = 0$ and (3.49) becomes

$$(I + \mu M)W + \sqrt{2}\,\mu'pQ = 0 \tag{3.52}$$

where $\mu' = \mu/pm$ is a dimensionless quantity. Substituting (3.39), (3.40), (3.42), and (3.43) into (3.52) gives

$$(EE^H + \mu E\Lambda E^H)EL + \sqrt{2}\,\mu'pEB = 0$$

or $\quad (EE^H E + \mu E\Lambda E^H E)L + \sqrt{2}\,\mu'pEB = 0$

or $\quad E[(1 + \mu\Lambda)]L + \sqrt{2}\,\mu'pEB] = 0$

or $\quad [(1 + \mu\Lambda)]L + \sqrt{2}\,\mu'pB] = 0 \tag{3.53}$

Hence, for the nth mode (3.51) gives

$$(1 + \mu\lambda_n)l_n + \sqrt{2}\,\mu'pb_n = 0 \tag{3.54a}$$

or $\quad l_{SSn} = -(\sqrt{2})\,\mu'p\dfrac{b_n}{1 + \mu\lambda_n} \tag{3.54b}$

where l_{nSS} is the steady-state values of the nth mode of the weighters and b_n is the nth mode of the cross-gain interference signal power between the main output port and all active AUX output ports.

Equation (3.54b) is identical to (2.72a) for a system with one interference signal source i and a large number of active AUX output ports. Thus, the subscript n in (3.54b) represents the nth interference signal source.

Therefore, when a matrix expression is decomposed into its modes, each mode corresponds to an interference signal source and hence the problem of a multisource, multioutput system reduces to that of a two-source, multioutput system as discussed in Section 2.4.

3.3.4 Feedback Loop

In Section 2.12, the feedback loop was analyzed for a single-loop system. In this section, equations for feedback loops in a multisource, multiport system are derived in terms of modal notations. In the complex frequency plane of $s = \sigma + j\omega$, the loop differential equation given by (3.51) will give

$$sW = -\frac{(I + \mu M)W}{T} - \frac{\mu\sqrt{2}}{mT}Q \tag{3.55}$$

Replacing M, Q, and W by their respective mode matrices, (3.55) becomes

$$sEL + \frac{(I + \mu E\Lambda E^H)}{T}EL = -\frac{\sqrt{2}\mu'p}{T}EB \tag{3.56}$$

Simplifying for the nth mode, (3.56) gives (3.54a)

$$sl_n + \frac{(1 + \mu\lambda_n)l_n}{T} = -\frac{\sqrt{2}\mu'p}{T}b_n$$

$$\text{or} \quad l_n\left[s + \frac{1 + \mu\lambda_n}{T}\right] = -\frac{\sqrt{2}\mu'}{T}v_n \tag{3.57}$$

where $v_n = pb_n$ and has the dimension of volt (2.213). Therefore, the loop transfer function H_n will be given by

$$H_n = \frac{l_n}{v_n} = -\frac{\sqrt{2}\mu'/T}{s + \dfrac{1 + \mu\lambda_n}{T}} = -\frac{\sqrt{2}\mu'/T}{s + s_n} \tag{3.58}$$

$$\text{where} \quad s_n = \frac{1 + \mu\lambda_n}{T} \tag{3.59}$$

which is identical to (2.216) where p_{Ai} has been replaced by λ_n. Therefore, all subsequent analyses of the feedback loop discussed in Section 2.12 will also be applicable here.

Note also that p_{Ai} is the interference signal power due to signal source i, appearing at all active AUX output ports, which is the AUX-A port. Therefore, if p_{Ai} is replaced by λ_n, then λ_n will represent the interference signal powers due to the signal source n, appearing at all active AUX output ports.

3.4 Physical Significance of Eigenvalues

Although expressions of eigenvalues have been derived above, it is not clear yet what they really are or what they stand for. Hence, the physical significance of eigenvalues is now discussed.

3.4.1 Signal Matrix in Terms of Signal Powers and Eigenvalues

From (3.31) and (3.38),

$$M = C^H R C = E \Lambda E^H \qquad (3.60)$$

Again, from (3.21) and (3.9),

$$C^H R C = C^H X X^H C = C^H S A A^H S^H C = \Delta P \Delta^H \qquad (3.61)$$

Hence, from (3.60) and (3.61),

$$M = E \Lambda E^H = \Delta P \Delta^H \qquad (3.62)$$

$$\text{where} \qquad \Delta = C^H S \qquad (3.63\text{a})$$

$$\Delta^H = S^H C \qquad (3.63\text{b})$$

$$P = A A^H = \begin{bmatrix} p_1 & 0 & 0 \\ 0 & p_m & 0 \\ 0 & 0 & p_{Ms} \end{bmatrix} \qquad (3.64)$$

where P represents the diagonal matrix of the signal powers associated with all transmitting interference signal sources and p_m is the power associated with the mth interference signal source.

The total number of interference signal sources is M_s whose maximum value, according to Section 3.3.2.3, is equal to the maximum number of active AUX output ports or $(N - 1)$.

Again, S represents the transfer functions or gains between all interference signal sources and all input ports of the beam former while C^H represents those between all input ports and all AUX output ports of the beam former. Hence, Δ will represent the transfer functions or gains between all interference signal sources and all AUX output ports (3.12a).

Now, the total number of AUX output ports is $(N-1)$. Hence, $E \Lambda E^H$ can be expressed as the series

$$E \Lambda E^H = \sum_{n=1}^{N-1} \lambda_n e_n e_n^H \qquad (3.65)$$

Similarly, the total number of interferences signal sources is Ms and hence $\Delta P \Delta^H$ can also be expressed as the summation

$$\Delta P \Delta^H = \sum_{m=1}^{Ms} p_m \Delta_m \Delta_m^H \tag{3.66}$$

Note that both λ_n and p_m are single-valued constant quantities and hence could be moved to the beginning of the matrix expressions.

Since the maximum number of Ms is $(N-1)$, (3.64) gives

$$\sum_{n=1}^{N-1} \lambda_n e_n e_n^H = \sum_{m=1}^{Ms} p_m \Delta_m \Delta_m^H = \sum_{m=1}^{N-1} p_m \Delta_m \Delta_m^H \tag{3.67}$$

which can now be used to establish a relationship between the eigenvalue λ and the interference signal power p.

3.4.2 Characteristics of the Transfer Function or Gain Matrix

The matrix representation Δ of the transfer function or gain was mentioned above. The characteristics of Δ are now discussed.

3.4.2.1 Expansion of Gain Matrix

Combining (3.4) and (3.12b) with (3.63a) gives

$$\Delta = C^H S = \begin{bmatrix} C_1^H \\ C_2^H \\ C_r^H \\ \cdots \end{bmatrix} |S_1 \quad S_2 \quad S_m \quad \cdots |$$

$$= \begin{bmatrix} C_1^H S_1 & C_1^H S_2 & C_1^H S_m & \cdots \\ C_2^H S_1 & C_2^H S_2 & C_r^H S_m & \cdots \\ C_r^H S_1 & C_r^H S_2 & C_r^H S_m & \cdots \\ \cdots & \cdots & \cdots & \cdots \end{bmatrix} \tag{3.68}$$

$$= \begin{bmatrix} \Delta_{11} & \Delta_{12} & \Delta_{1m} & \cdots \\ \Delta_{21} & \Delta_{22} & \Delta_{2m} & \cdots \\ \Delta_{r1} & \Delta_{r2} & \Delta_{rm} & \cdots \\ \cdots & \cdots & \cdots & \cdots \end{bmatrix} = |\Delta_1 \quad \Delta_2 \quad \Delta_m \quad \cdots |$$

$$\text{where} \quad \Delta_m = \begin{bmatrix} \Delta_{1m} \\ \Delta_{2m} \\ \Delta_{rm} \\ \cdots \end{bmatrix} = \begin{bmatrix} C_1^H S_m \\ C_2^H S_m \\ C_r^H S_m \\ \cdots \end{bmatrix} \qquad (3.69)$$

Thus, Δ_m represents the gain vector due to signal source m and corresponds to the gains between signal source m and all AUX output ports.

Similarly,

$$\Delta^H = S^H C = \begin{bmatrix} S_1^H \\ S_2^H \\ S_m^H \\ \cdots \end{bmatrix} | C_1 \quad C_2 \quad C_r \quad \cdots |$$

$$= \begin{bmatrix} S_1^H C_1 & S_1^H C_2 & S_1^H C_r & \cdots \\ S_2^H C_1 & S_2^H C_2 & S_2^H C_r & \cdots \\ S_m^H C_1 & S_m^H C_2 & S_m^H C_r & \cdots \\ \cdots & \cdots & \cdots & \cdots \end{bmatrix} \qquad (3.70)$$

$$= \begin{bmatrix} \Delta_{11}^H & \Delta_{22}^H & \Delta_{r1}^H & \cdots \\ \Delta_{12}^H & \Delta_{22}^H & \Delta_{r2}^H & \cdots \\ \Delta_{1m}^H & \Delta_{2m}^H & \Delta_{rm}^H & \cdots \\ \cdots & \cdots & \cdots & \cdots \end{bmatrix} = \begin{bmatrix} \Delta_1^H \\ \Delta_2^H \\ \Delta_m^H \\ \cdots \end{bmatrix}$$

$$\text{where} \quad \Delta_m^H = | \Delta_{1m}^H \quad \Delta_{2m}^H \quad \Delta_{rm}^H \quad \cdots | \qquad (3.71)$$
$$= | S_m^H C_1 \quad S_m^H C_2 \quad S_m^H C_r \quad \cdots |$$

Note that Δ_m^H is the Hermitian representation of the gain vector Δ_m. Therefore,

$\Delta_{rm} = | G_r(\theta_m)$ = transfer function or gain between the mth interference signal source and the rth AUX output port.

Δ_m = transfer function or gain between the mth interference signal source and all AUX output ports. It is represented by a column matrix, or column vector, as given by (3.69) where the number of elements in the column matrix is equal to the number of the active AUX output ports.

Δ = transfer function or gain between all interference signal sources and all AUX output ports. It is represented by a rectangular matrix where the number of columns is equal to the number of transmitting interference signal sources and the number of rows is equal to the number of active AUX output ports.

3.4.2.2 Gain Vectors for Signal Sources Along Arbitrary Directions

Consider the mth interference signal source located along an arbitrary direction of $+\theta_m$ off the array boresight. The corresponding gain vector, Δ_m, is given by (3.69) and its Hermitian Δ_m^H is given by (3.71). Therefore, the product $\Delta_m^H \Delta_m$ is given by

$$\Delta_m^H \Delta_m = |S_m^H C_1 \quad S_m^H C_2 \quad S_m^H C_r \quad \ldots| \begin{bmatrix} C_1^H S_m \\ C_2^H S_m \\ C_r^H S_m \\ \ldots \end{bmatrix} \quad (3.72)$$

$$= |S_m^H C_1 C_1^H S_m + S_m^H C_2 C_2^H S_m + S_m^H C_r C_r^H S_m + \ldots$$

Now, the beam former has N input ports and so the signal from the mth interference signal source will reach the rth AUX output port through all N input ports of the beam former. Hence,

$$C_r^H S_m = |C_{r1} \quad C_{r2} \quad C_{rn} \quad C_{rN}|^* \begin{bmatrix} S_{1m} \\ S_{2m} \\ S_{nm} \\ S_{Nm} \end{bmatrix}$$

$$= C_{r1}^* S_{1m} + C_{r2}^* S_{2m} + \ldots + C_{rn}^* S_{nm} + \ldots + C_{rN}^* S_{Nm}$$

$$= |G_{r1}(\theta_m)| + |G_{r2}(\theta_m)| + \ldots + |G_{rn}(\theta_m)| + \ldots \quad (3.73)$$

$$+ |G_{rN}(\theta_m)|$$

$$= |G_r(\theta_m)|$$

where

$|G_{rn}(\theta_m)|$ = the transfer function or gain between the mth interference signal source and the rth AUX output port through the nth input port of the beam former;

$|G_r(\theta_m)|$ = the transfer function or gain between the mth interference signal source and the rth AUX output port through all input ports of the beam former.

Similarly,

$$S_m^H C_r = |S_{1m} \quad S_{2m} \quad S_{nm} \quad S_{Nm}|^* \begin{bmatrix} C_{r1} \\ C_{r2} \\ C_{rn} \\ C_{rN} \end{bmatrix}$$

$$= S_{1m}^* C_{r1} + S_{2m}^* C_{r2} + \ldots + S_{nm}^* C_{rn} + \ldots + S_{Nm}^* C_{rN}$$

$$= |G_{r1}(\theta_m)| + |G_{r2}(\theta_m)| + \ldots + |G_{rn}(\theta_m)| + \ldots \quad (3.74)$$
$$+ |G_{rN}(\theta_m)|$$

$$= |G_r(\theta_m)|$$

Combining (3.73) and (3.74) gives

$$S_m^H C_r C_r^H S_m = |G_r(\theta_m)|^2 \qquad (3.75)$$

Therefore, (3.72) gives

$$\Delta_m^H \Delta_m = |G_1(\theta_m)|^2 + |G_2(\theta_m)|^2 + \ldots + |G_2(\theta_m)|^2 + \ldots \qquad (3.76)$$

Thus, the product of the Hermitian of a gain vector and the gain vector itself, due to the mth interference signal source, located along an arbitrary direction of $+\theta_m$ off the array boresight, is the sum of the square of the transfer functions or gains between the signal source and all AUX output ports.

3.4.2.3 Gain Vectors for Signal Sources Along AUX Beams

Let two interference signal sources be located along the peaks of the AUX beams n and r of an orthogonal beam former.

Now, as discussed in Chapter 1, if a signal source is placed along the peak of the nth beam, located at $+\theta_n$ off the boresight, then the signal will appear at the nth AUX output port only. Similarly, if a signal source is placed along the peak of the rth beam, located at $+\theta_r$ off the boresight, then the signal will appear at the rth AUX output port only.

As mentioned above, the gain vector Δ_n represents the gains between the signal source n and all AUX output ports, and the gain vector Δ_r represents that between the signal source r and all AUX output ports. However, in the present case, the signal sources n and r are located along the AUX beams n and r, respectively. Therefore, all elements of the gain vector Δ_n, due to a signal source located along the AUX beam n, will be zero, except the element Δ_{nn}. Similarly, all elements of the gain vector Δ_r, due to a signal source located along the AUX beam r, will be zero, except the element Δ_{rr}. Hence, the vectors Δ_n and Δ_r are given by

$$\Delta_n = \begin{bmatrix} \Delta_{1n} \\ \Delta_{2n} \\ \Delta_{nn} \\ \Delta_{rn} \\ \Delta_{(N-1)n} \end{bmatrix} = \begin{bmatrix} 0 \\ 0 \\ \Delta_{nn} \\ 0 \\ 0 \end{bmatrix} \qquad \Delta_r = \begin{bmatrix} \Delta_{1r} \\ \Delta_{2r} \\ \Delta_{nr} \\ \Delta_{rr} \\ \Delta_{(N-1)r} \end{bmatrix} = \begin{bmatrix} 0 \\ 0 \\ 0 \\ \Delta_{rr} \\ 0 \end{bmatrix} \tag{3.77}$$

Therefore, $\Delta_n^H \Delta_r = \Delta_r^H \Delta_n = 0$ (3.78)

Hence, the product of the Hermitian of the gain vector due to one signal source and the gain vector due to another signal source will become zero if the two signal sources are located along two different AUX beams of an orthogonal beam former.

3.4.3 Relation Between Eigenvalues and Signal Powers

As has been mentioned in connection with (3.57), λ_n represents the interference signal powers at the AUX output ports. However, the exact relationship between the AUX port signal powers and λ_n is still unknown.

Now, in (3.67), the total number of terms contained in both left-hand side series and right-hand side series summations are the same because they both are $(N-1)$. Also, both p_m and λ_n have the dimension of power. Therefore, since the vector Δ_m represents transfer function or gain, the vector e_n will also represent the same. Therefore, to establish a relationship between the signal power and the eigenvalue, the eigen vector e_n in (3.35) will be expressed as a linear combination of the gain vector Δ in (3.69).

Thus, for transmitting interference signal sources, 1, 2, . . . , m, . . . , the sum e of the corresponding eigenvectors can be expressed as

$$e = b_1\Delta_1 + b_2\Delta_2 + \ldots + b_m\Delta_m + \ldots + b_{(N-1)}\Delta_{(N-1)} \quad (3.79)$$

where b_1, b_2, ..., are constants and since the maximum number of signal sources is $(N-1)$, the vector $\Delta_{(N-1)}$ will represent the gains between the interference signal source $(N-1)$ and all AUX output ports.

Again, expanding the right-hand side of (3.67) into a series corresponding to the number of interference signal sources and using (3.60) gives

$$M = p_1\Delta_1\Delta_1^H + p_2\Delta_2\Delta_2^H + \ldots + p_m\Delta_m\Delta_m^H + \ldots \quad (3.80)$$
$$+ p_{(N-1)}\Delta_{(N-1)}\Delta_{(N-1)}^H$$

Therefore, from (3.34)

$$Me = e\lambda$$

or

$$\{p_1\Delta_1\Delta_1^H + p_2\Delta_2\Delta_2^H + \ldots + p_m\Delta_m\Delta_m^H + \ldots + p_{(N-1)}\Delta_{(N-1)}\Delta_{(N-1)}^H]$$
$$\times [b_1\Delta_1 + b_2\Delta_2 + \ldots + b_m\Delta_m + \ldots + b_{(N-1)}\Delta_{(N-1)}]$$
$$= [b_1\Delta_1 + b_2\Delta_2 + \ldots + b_m\Delta_m + \ldots + b_{(N-1)}\Delta_{(N-1)}]\lambda \quad (3.81)$$

which can now be solved to establish the desired relationship between the signal powers and the eigenvalues.

Example 3.1

An adaptive system contains a single interference signal source m located at $+\theta_m$ off the array boresight. Determine the corresponding eigenvalues in terms of the interference signal power.

Answer

For the single interference source m, (3.81) becomes

$$[p_m\Delta_m\Delta_m^H][b_m\Delta_m] = [b_m\Delta_m]\lambda$$

$$\text{or} \quad b_m p_m \Delta_m \Delta_m^H \Delta_m = b_m \lambda \Delta_m \quad (3.82a)$$

$$\text{Put} \quad \Delta_m^H \Delta_m = f_m \quad (3.82b)$$

$$\text{Then} \quad b_m p_m \Delta_m f_m = b_m \lambda \Delta_m$$

$$\text{or} \quad b_m p_m f_m \Delta_m = b_m \lambda \Delta_m$$

$$\text{or} \quad p_m f_m = \lambda = \lambda_m \tag{3.82c}$$

Therefore, combining (3.76), (3.82b), and (3.82c) gives

$$\lambda_m = [|G_1(\theta_m)|^2 + |G_2(\theta_m)|^2 + \ldots + |G_r(\theta_m)|^2 + \ldots] p_m \tag{3.83}$$

$$= p_{1m} + p_{2m} + \ldots + p_{rm} + \ldots$$

Thus, the system has a single eigenvalue which is equal to the sum of the signal powers at the outputs of all active AUX ports. Note that (3.83) is identical to (2.67) and hence the null depths derived in Section 2.4.3 are, in fact, in terms of the system eigenvalues.

Example 3.2

An adaptive system has two interference signal sources m and n located at $+\theta_m$ and $+\theta_n$ off the array boresight, respectively. Determine the corresponding eigenvalues in terms of the interference signal powers.

Answer

For two interference signal sources, m and n, (3.81) becomes

$$[p_m \Delta_m \Delta_m^H + p_n \Delta_n \Delta_n^H][b_m \Delta_m + b_n \Delta_n] = [b_m \Delta_m + b_n \Delta_n]\lambda$$

$$\text{or} \quad b_m p_m \Delta_m \Delta_m^H \Delta_m + b_n p_m \Delta_m \Delta_m^H \Delta_n + b_m p_n \Delta_n \Delta_n^H \Delta_m$$

$$+ b_n p_n \Delta_n \Delta_n^H \Delta_n \tag{3.84}$$

$$= b_m \Delta_m \lambda + b_n \Delta_n \lambda$$

Now, let θ be the angle between Δ_m and Δ_n and put

$$\phi = \cos\theta \tag{3.85a}$$

$$\text{Also, let} \quad f_m = \Delta_m^H \Delta_m \tag{3.85b}$$

$$\text{And} \quad f_n = \Delta_n^H \Delta_n \tag{3.85c}$$

Then $\Delta_m^H \Delta_n$ and $\Delta_n^H \Delta_m$ can be expressed as

$$\Delta_m^H \Delta_n = [\sqrt{(f_m f_n)}]\phi = k\phi \tag{3.86a}$$

$$\Delta_n^H \Delta_m = [\sqrt{(f_m f_n)}]\phi^* = k\phi^* \tag{3.86b}$$

$$\text{where} \quad k = \sqrt{(f_m f_n)} \tag{3.86c}$$

Then, (3.84) can be written

$$[b_m p_m f_m + b_n p_m k\phi]\Delta_m \tag{3.87}$$
$$+ [b_n p_n f_n + b_m p_n k\phi^*]\Delta_n = [b_m \lambda]\Delta_m + [b_n \lambda]\Delta_n$$

Represented in matrix form, (3.87) becomes

$$\begin{bmatrix} p_m f_m & p_m k\phi \\ p_n k\phi^* & p_n f_n \end{bmatrix} \begin{bmatrix} b_m \\ b_n \end{bmatrix} = \lambda \begin{bmatrix} b_m \\ b_n \end{bmatrix}$$

$$\text{or} \quad \begin{bmatrix} p_m f_m & p_m k\phi \\ p_n k\phi^* & p_n f_n \end{bmatrix} = \lambda I \tag{3.88}$$

where I is a unit matrix (8, Appendix I). Put

$$\begin{bmatrix} p_m f_m & p_m k\phi \\ p_n k\phi^* & p_n f_n \end{bmatrix} = M' \tag{3.89}$$

Then (3.88) reduces to

$$|M' - \lambda I| = 0 \tag{3.90}$$

which is identical to (3.32) and, hence, λ can be obtained from the determinant of (3.90) (11, Appendix I). Thus,

$$\det|M' - \lambda I| = 0$$

$$\text{or} \quad \det \begin{bmatrix} p_m f_m - \lambda & p_m k\phi \\ p_n k\phi^* & p_n f_n - \lambda \end{bmatrix} = 0$$

$$\text{or} \quad \lambda^2 - [p_m f_m + p_n f_n]\lambda + p_m p_n k^2 [1 - |\phi|^2] = 0 \tag{3.91}$$

Hence,

$$\lambda = \frac{p_m f_m + p_n f_n}{2} \pm \frac{p_m f_m + p_n f_n}{2} \sqrt{1 - \frac{4 p_m p_n k^2 [1 - |\phi|^2]}{(p_m f_m + p_n f_n)^2}}$$

(3.92)

Thus, two interference signal sources will correspond to two eigenvalues. The magnitudes of these eigenvalues will depend on the angle θ between the two gain vectors or the Δ vectors. Two special cases will be considered here where (a) the two Δ vectors are orthogonal (i.e., $\theta = 90°$) and (b) the two Δ vectors are parallel ($\theta = 0$).

(a) Orthogonal Δ Vectors

In this case the two signal sources are located along the peaks of two AUX beams m and n. Since the Δ vectors are orthogonal, $\theta = 90°$ and hence, $\phi = 0$. Therefore, (3.52) gives

$$\lambda = \frac{p_m f_m + p_n f_n}{2} \pm \frac{p_m f_m - p_n f_n}{2}$$

(3.93)

The two resulting eigenvalues will therefore be given by

$$\lambda_m = p_m f_m$$
$$= [|G_1(\theta_m)|^2 + |G_2(\theta_m)|^2 + \ldots + |G_r(\theta_m)|^2 + \ldots] p_m \quad (3.94a)$$
$$= p_{1m} + p_{2m} + \ldots + p_{rm} + \ldots$$

is the sum of the signal powers appearing at all active AUX output ports due to the interference signal source m.

$$\lambda_n = p_n f_n$$
$$= [|G_1(\theta_n)|^2 + |G_2(\theta_n)|^2 + \ldots + |G_r(\theta_n)|^2 + \ldots] p_n \quad (3.94b)$$
$$= p_{1n} + p_{2n} + \ldots + p_{rn} + \ldots$$

is the sum of the signal powers appearing at all active AUX output ports due to the interference signal source n.

(b) Parallel Δ Vectors

In this case the two interference signal sources m and n are located along the same AUX beam, say, beam m.

Since the Δ vectors are parallel, $\theta = 0$ and $\phi = 1$. Therefore, (3.92) gives

$$\lambda = \frac{p_m f_m + p_n f_n}{2} \pm \frac{p_m f_m + p_n f_n}{2} \qquad (3.95)$$

Therefore, the two resulting eigenvalues will be given by

$$\lambda_m = p_m f_m + p_n f_n \qquad (3.96a)$$
$$= [p_{1m} + p_{2m} + \ldots + p_{rm} + \ldots]$$
$$+ [p_{1n} + p_{2n} + \ldots + p_{rn} + \ldots] \qquad (3.96b)$$
$$\lambda_n = 0$$

Thus, when two signal sources are colocated, they could be regarded as a single source transmitting two different signals. Hence, the two eigenvalues will become one and will contain the sum of the signal powers at all active AUX output ports due to both signal sources m and n.

Example 3.3

An adaptive system has a large number of associated interference signal sources. Determine the corresponding eigenvalues.

Answer

(a) If all related gain vectors are mutually orthogonal, then the number of eigenvalues will be equal to the number of interference signal sources. Thus,

$$\lambda_1 = p_1 f_1 = p_{11} + p_{21} + \ldots + p_{r1} + \ldots$$
$$\lambda_2 = p_2 f_2 = p_{12} + p_{22} + \ldots + p_{r2} + \ldots$$
$$\ldots$$
$$\lambda_m = p_m f_m + p_{1m} + p_{2m} + \ldots + p_{rm} + \ldots$$
$$\ldots$$

(b) If all gain vectors are parallel, all eigenvalues will become a single one given by

$$\lambda = p_1 f_1 + p_2 f_2 + \ldots + p_m f_m + \ldots$$
$$= [p_{11} + p_{21} + \ldots + p_{r1} + \ldots]$$
$$+ [p_{12} + p_{22} + \ldots + p_{r2} + \ldots]$$
$$\ldots$$
$$+ [p_{1m} + p_{2m} + \ldots + p_{rm} + \ldots]$$

References

[1]   Hudson, J. E., *Adaptive Array Principles,* London: Institution of Electrical Engineers, 1981.

[2]   Klemes, M., "A Practical Method of Obtaining Constant Convergence Rate in LMS Adaptive Arrays," *IEEE Trans. on Antennas and Propagation,* Vol. AP-34, March 1986, pp. 440–446.

[3]   Crompton, Jr., R. T., "Improved Feedback Loop for Adaptive Arrays," *IEEE Trans. on Aerospace and Electronic Systems,* Vol. AES-16, March 1980, pp. 159–168.

[4]   Selby, S. M. (ed.), *Standard Mathematical Tables,* Chemical Rubber Company: Cleveland, OH, 1969, pp. 118–140.

Appendix **A**

Final Output Power for a Two-Source, Two-Output System

From (2.32), the final steady-state output power p_{oSS}, at the output port of the signal processing network, is given by

$$4p_{oSS} = \frac{m^2 W_A^2}{2} |G_A(\theta_i)|^2 p_i + |G_M(0)|^2 p_d + |G_M(\theta_i)|^2 p_i \quad \text{(A.1)}$$

$$+ \frac{2mW_A}{\sqrt{2}} |G_M(\theta_i) G_A(\theta_i)| p_i$$

At steady state, $W_A = W_{ASS}$ and hence, rearranging (A.1) gives

$$4p_{oSS} = |G_M(0)|^2 p_d + |G_M(\theta_i)|^2 p_i + \frac{m^2 W_{ASS}^2}{2} |G_A(\theta_i)|^2 p_i \quad \text{(A.2)}$$

$$+ \frac{2mW_{ASS}}{\sqrt{2}} |G_M(\theta_i) G_A(\theta_i)| p_i$$

Substitution of W_{ASS} from (2.27) into (A.2) gives

229

$$4p_{oSS} = |G_M(0)|^2 p_d + |G_M(\theta_i)|^2 p_i$$

$$+ \frac{m^2}{2} \left[\frac{\sqrt{2}}{m} \frac{|G_A(\theta_i)G_M(\theta_i)|}{(1/\mu) + |G_A(\theta_i)|^2 p_i} \right]^2 |G_A(\theta_i)|^2 p_i$$

$$+ \frac{2m}{\sqrt{2}} \left[-\frac{m}{\sqrt{2}} \frac{|G_A(\theta_i)G_M(\theta_i)| p_i}{(1/\mu) + |G_A(\theta_i)|^2 p_i} \right] |G_A(\theta_i)G_M(\theta_i)| p_i \qquad (A.3)$$

$$= |G_M(0)|^2 p_d + |G_M(\theta_i)|^2 p_i + \left[\frac{|G_A(\theta_i)G_M(\theta_i)|}{(1/\mu) + |G_A(\theta_i)|^2 p_i} \right]^2 |G_A(\theta_i)|^2 p_i$$

$$- 2 \left[\frac{|G_A(\theta_i)G_M(\theta_i)| p_i}{(1/\mu) + |G_A(\theta_i)|^2 p_i} \right] |G_A(\theta_i)G_M(\theta_i)| p_i$$

Rearranging the gain functions, (A.3) gives

$$4p_{oSS} = |G_M(0)|^2 p_d + |G_M(\theta_i)|^2 p_i$$

$$+ \frac{[|G_A(\theta_i)|^2 p_i]}{[(1/\mu) + |G_A(\theta_i)|^2 p_i]^2} |G_M(\theta_i)|^2 p_i \qquad (A.4)$$

$$- 2 \frac{|G_A(\theta_i)|^2 p_i}{(1/\mu) + |G_A(\theta_i)|^2 p_i} |G_M(\theta_i)|^2 p_i$$

Put $|G_M(0)|^2 p_d = p_{Md}$ = desired signal power at the main output port

$|G_M(\theta_i)|^2 p_i = p_{Mi}$ = interference signal power at the main output port

$|G_A(\theta_i)|^2 p_i = p_{Ai}$ = interference signal power at the AUX-A output port

Then, (A.4) becomes

$$4p_{oSS} = p_{Md} + p_{Mi} + \frac{[p_{Ai}]^2}{[(1/\mu) + p_{Ai}]^2} p_{Mi} - 2 \frac{p_{Ai}}{(1/\mu) + p_{Ai}} p_{Mi}$$

$$= p_{Md} + p_{Mi} \left[1 + \frac{[p_{Ai}]^2}{[(1/\mu) + p_{Ai}]^2} - 2 \frac{p_{Ai}}{(1/\mu) + p_{Ai}} \right] \qquad (A.5)$$

$$= p_{Md} + p_{Mi} \left[1 - \frac{p_{Ai}}{(1/\mu) + p_{Ai}} \right]^2 = p_{Md} + \frac{p_{Mi}}{[(1/\mu) + p_{Ai}]^2}$$

This is the same as (2.33a) and represents the final output power as a sum of the main port desired power and the main port interference power.

Appendix B

Final Output Power for a Two-Source, Three-Output System

From (2.52), the final steady-state output power p_{oSS}, at the output port of the signal processing network, is given by

$$
\begin{aligned}
(2N)p_{oSS} &= \left[\frac{mW_{SS1}\,y_1(t)}{\sqrt{2}} + \frac{mW_{SS2}\,y_2(t)}{\sqrt{2}} + y_M(t) \right]^2 \\
&= \frac{m^2 W_{SS1}^2 |y_1(t)|^2}{2} + \frac{m^2 W_{SS2}^2 |y_2(t)|^2}{2} + |y_M(t)|^2 \qquad \text{(B.1)} \\
&\quad + \frac{2m^2 W_{SS1} W_{SS2} |y_1(t)\,y_2(t)|}{2} \\
&\quad + \frac{2m\,\mathrm{Re}\,W_{SS1}|y_1(t)\,y_M(t)|}{\sqrt{2}} + \frac{2m\,\mathrm{Re}\,W_{SS2}|y_2(t)\,y_M(t)|}{\sqrt{2}}
\end{aligned}
$$

Using (2.35) and (2.36),

$$
\begin{aligned}
|y_1(t)|^2 &= |G_1(\theta_i)|^2 p_i = p_{1i} \\
|y_2(t)|^2 &= |G_2(\theta_i)|^2 p_i = p_{2i} \\
|y_M(t)|^2 &= |G_M(0)|^2 p_d + |G_M(\theta_i)|^2 p_i = p_{Md} + p_{Mi} \\
|y_1(t)\,y_2(t)| &= |G_1(\theta_i)\,G_2(\theta_i)|\,p_i \\
|y_1(t)\,y_M(t)| &= |G_1(\theta_i)\,G_M(\theta_i)|\,p_i \\
|y_2(t)\,y_M(t)| &= |G_2(\theta_i)\,G_M(\theta_i)|\,p_i
\end{aligned}
$$

where $p_i = |f_i(t)|^2$ = the interference signal power and $p_d = |f_d(t)|^2$ = the desired signal power. Therefore, (B.1) becomes

$$
\begin{aligned}
(2N)p_{oSS} = {}& \frac{m^2 W_{SS1}^2 |G_1(\theta_i)|^2 p_i}{2} + \frac{m^2 W_{SS2}^2 |G_2(\theta_i)|^2 p_i}{2} \\
&+ |G_M(0)|^2 p_d + |G_M(\theta_i)|^2 p_i \\
&+ \frac{2m^2 W_{SS1} W_{SS2} |G_1(\theta_i) G_2(\theta_i)| p_i}{2} \\
&+ \frac{2m W_{SS1} |G_1(\theta_i) G_M(\theta_i)| p_i}{\sqrt{2}} + \frac{2m W_{SS2} |G_2(\theta_i) G_M(\theta_i)| p_i}{\sqrt{2}}
\end{aligned}
\tag{B.2}
$$

Now, in (2.49) put

$$
1 + \mu p_{1i} + \mu p_{2i} = K
\tag{B.3}
$$

Then the steady-state weighter values will be given by

$$
W_{SS1} = -\frac{(\sqrt{2})\mu |G_1(\theta_i) G_M(\theta_i)| p_i}{mK}
\tag{B.4a}
$$

$$
W_{SS2} = -\frac{(\sqrt{2})\mu |G_2(\theta_i) G_M(\theta_i)| p_i}{mK}
\tag{B.4b}
$$

Substitution of (B.4) into (B.2) gives

$$
\begin{aligned}
(2N)p_{oSS} = {}& \frac{[\mu |G_1(\theta_i) G_M(\theta_i)| p_i]^2 |G_1(\theta_i)|^2 p_i}{K^2} \\
&+ \frac{[\mu |G_2(\theta_i) G_M(\theta_i)| p_i]^2 |G_2(\theta_i)|^2 p_i}{K^2} \\
&+ |G_M(0)|^2 p_d + |G_M(\theta_i)|^2 p_i \\
&+ \frac{2\mu^2 [|G_1(\theta_i) G_M(\theta_i)| p_i][|G_2(\theta_i) G_M(\theta_i)| p_i] |G_1(\theta_i) G_2(\theta_i)| p_i}{K^2} \\
&- \frac{2\mu |G_1(\theta_i) G_M(\theta_i)| p_i |G_1(\theta_i) G_M(\theta_i)| p_i}{K} \\
&- \frac{2\mu [|G_2(\theta_i) G_M(\theta_i)| p_i] |G_2(\theta_i) G_M(\theta_i)| p_i}{K}
\end{aligned}
\tag{B.5}
$$

Rearranging the gain functions in (B.5) gives

$$
(2N)p_{oSS} = \frac{(\mu |G_1(\theta_i)|^2 p_i)^2 |G_M(\theta_i)|^2 p_i}{K^2}
$$

$$
+ \frac{[\mu |G_2(\theta_i)|^2 p_i]^2 |G_M(\theta_i)|^2 p_i}{K^2}
$$

$$
+ |G_M(0)|^2 p_d + |G_M(\theta_i)|^2 p_i \qquad \text{(B.6)}
$$

$$
+ \frac{2\mu^2 [|G_1(\theta_i)|^2 p_i][|G_2(\theta_i)|^2 p_i]}{K^2} |G_M(\theta_i)|^2 p_i
$$

$$
- \frac{2\mu |G_1(\theta_i)|^2 p_i}{K} |G_M(\theta_i)|^2 p_i
$$

$$
- \frac{2\mu (|G_2(\theta_i)|^2 p_i)}{K} |G_M(\theta_i)|^2 p_i
$$

As before,

$|G_M(0)|^2 p_d = p_{Md}$ = the desired signal power at the main output port;

$|G_M(\theta_i)|^2 p_i = p_{Mi}$ = the interference signal power at the main output port;

$|G_1(\theta_i)|^2 p_i = p_{1i}$ = the interference signal power at the AUX-1 output port;

$|G_2(\theta_i)|^2 p_i = p_{2i}$ = the interference signal power at the AUX-2 output port.

Then, (B.6) becomes

$$
(2N)p_{oSS} = p_{Md} + p_{Mi} + \frac{[\mu p_{1i}]^2}{K^2} p_{Mi} + \frac{[\mu p_{2i}]^2}{K^2} p_{Mi} + \frac{2[\mu^2 p_{1i} p_{2i}]}{K^2} p_{Mi}
$$

$$
- \frac{2[\mu p_{1i}]}{K} p_{Mi} - \frac{2[\mu p_{2i}]}{K} p_{Mi}
$$

$$
= p_{Md} + p_{Mi} \left[1 + \frac{[\mu p_{1i}]^2}{K^2} + \frac{[\mu p_{2i}]^2}{K^2} + \frac{2[\mu^2 p_{1i} p_{2i}]}{K^2} - \frac{2[\mu p_{1i}]}{K} - \frac{2[\mu p_{2i}]}{K} \right]
$$

$$= p_{Md} + p_{Mi} \left[1 - \frac{\mu p_{1i}}{K} - \frac{\mu p_{2i}}{K} \right]^2$$

$$= p_{Md} + p_{Mi} \left[1 - \frac{\mu p_{1i}}{1 + \mu p_{1i} + \mu p_{2i}} - \frac{\mu p_{2i}}{1 + \mu p_{1i} + \mu p_{2i}} \right]^2$$

$$= p_{Md} + \frac{p_{Mi}}{[1 + \mu p_{1i} + \mu p_{2i}]^2} \tag{B.7}$$

Appendix C

Final Output Power for a Three-Source, Two-Output System

From (2.93), the final steady-state output power p_{oSS}, at the output port of the signal processing network, is given by

$$
\begin{aligned}
4p_{oSS} &= \left[\frac{m W_{aSS} y_a(t)}{\sqrt{2}} + y_M(t) \right]^2 \\
&= \frac{m^2 W_{aSS}^2 |y_a(t)|^2}{2} + |y_M(t)|^2 + \frac{2m \operatorname{Re} W_{aSS} |y_a(t) y_M(t)|}{\sqrt{2}}
\end{aligned}
\tag{C.1}
$$

where

$$
\begin{aligned}
|y_a(t)|^2 &= |G_a(\theta_1)|^2 p_1 + |G_a(\theta_2)|^2 p_2 = p_{a1} + p_{a2}; \\
|y_M(t)|^2 &= |G_M(0)|^2 p_d + |G_M(\theta_1)|^2 p_1 + |G_M(\theta_2)|^2 p_2 \\
&= p_{Md} + p_{M1} + p_{M2};
\end{aligned}
$$

and $|y_a(t) y_M(t)| = |G_a(\theta_1) G_M(\theta_1)| p_1 + |G_a(\theta_2) G_M(\theta_2)| p_2$.

From (2.93b), the steady-state value of the weighter W_{aSS} is

$$
W_{aSS} = - \frac{(\sqrt{2}) \mu [|G_a(\theta_1) G_M(\theta_1)| p_1 + |G_a(\theta_2) G_M(\theta_2)| p_2]}{mK}
\tag{C.2}
$$

$$\text{where} \quad K = 1 + \mu p_{a1} + \mu p_{a2} \tag{C.3}$$

Substitution of (C.2) into (C.1) gives

$$4 p_{oSS} = \frac{2m^2}{2m^2} \frac{\mu^2 [|G_a(\theta_1)G_M(\theta_1)| p_1 + |G_a(\theta_2)G_M(\theta_2)| p_2]^2}{K^2} \tag{C.4}$$

$$\times [|G_a(\theta_1)|^2 p_1 + |G_a(\theta_1)|^2 p_1]$$

$$+ |G_M(0)|^2 p_d + |G_M(\theta_1)|^2 p_1 + |G_M(\theta_2)|^2 p_2$$

$$- \frac{(\sqrt{2})m}{(\sqrt{2})m} \frac{\mu[|G_a(\theta_1)G_M(\theta_1)| p_1 + |G_a(\theta_2)G_M(\theta_2)| p_2]}{K}$$

$$\times [|G_a(\theta_1)G_M(\theta_1)| p_1 + |G_a(\theta_2)G_M(\theta_2)| p_2]$$

$$= p_{Md} + p_{M1} + p_{M2}$$

$$+ \frac{\mu^2 [|G_a(\theta_1)G_M(\theta_1)| p_1 + |G_a(\theta_2)G_M(\theta_2)| p_2]^2}{K^2} [|G_a(\theta_1)|^2 p_1]$$

$$+ \frac{\mu^2 [|G_a(\theta_1)G_M(\theta_1)| p_1 + |G_a(\theta_2)G_M(\theta_2)| p_2]^2}{K^2} [|G_a(\theta_2)|^2 p_2]$$

$$+ \frac{\mu[|G_a(\theta_1)G_M(\theta_1)| p_1 + |G_a(\theta_2)G_M(\theta_2)| p_2]}{K} [|G_a(\theta_1)G_M(\theta_1)| p_1]$$

$$+ \frac{\mu[|G_a(\theta_1)G_M(\theta_1)| p_1 + |G_a(\theta_2)G_M(\theta_2)| p_2]}{K} [|G_a(\theta_2)G_M(\theta_2)| p_2]$$

Rearranging the gain functions, (C.4) gives

$$4 p_{oSS} = p_{Md} + p_{M1} + p_{M2} \tag{C.5}$$

$$+ \frac{\mu^2}{K^2} \left[|G_a(\theta_1)|^2 p_1 + \frac{|G_a(\theta_1)/G_a(\theta_2)|}{|G_M(\theta_1)/G_M(\theta_2)|} |G_a(\theta_2)|^2 p_2 \right]^2 [|G_M(\theta_1)|^2 p_1]$$

$$+ \frac{\mu^2}{K^2} \left[\frac{|G_a(\theta_2)/G_a(\theta_1)|}{|G_M(\theta_2)/G_M(\theta_1)|} |G_a(\theta_1)|^2 p_1 + |G_a(\theta_2)|^2 p_2 \right]^2 [|G_M(\theta_2)|^2 p_2]$$

$$- \frac{2\mu}{K} \left[|G_a(\theta_1)|^2 p_1 + \frac{|G_a(\theta_1)/G_a(\theta_2)|}{|G_M(\theta_1)/G_M(\theta_2)|} |G_a(\theta_2)|^2 p_2 \right] [|G_M(\theta_1)|^2 p_1]$$

$$- \frac{2\mu}{K} \left[\frac{|G_a(\theta_2)/G_a(\theta_1)|}{|G_M(\theta_2)/G_M(\theta_1)|} |G_a(\theta_1)|^2 p_1 + |G_a(\theta_2)|^2 p_2 \right]^2 [|G_M(\theta_2)|^2 p_2]$$

$$\text{Put} \quad K_{a12} = \frac{1}{K_{a21}} = \frac{|G_a(\theta_1)/G_a(\theta_2)|}{|G_M(\theta_1)/G_M(\theta_2)|} \tag{C.6}$$

Then, (C.5) becomes

$$4p_{oSS} = p_{Md} + p_{M1} + p_{M2} + \frac{\mu^2}{K^2}[p_{a1} + K_{a12}p_{a2}]^2 p_{M1}$$

$$+ \frac{\mu^2}{K^2}[K_{a21}p_{a1} + p_{a2}]^2 p_{M2} - 2\frac{\mu}{K}[p_{a1} + K_{a12}p_{a2}]p_{M1}$$

$$- 2\frac{\mu}{K}[K_{a21}p_{a1} + p_{a2}]p_{M2} \tag{C.7}$$

$$= p_{Md} + p_{M1}\left[1 + \frac{\mu^2}{K^2}[p_{a1} + K_{a12}p_{a2}]^2 - 2\frac{\mu}{K}[p_{a1} + K_{a12}p_{a2}]\right]$$

$$+ p_{M2}\left[1 + \frac{\mu^2}{K^2}[K_{a21}p_{a1} + p_{a2}]^2 - 2\frac{\mu}{K}[K_{a21}p_{a1} + p_{a2}]\right]$$

$$= p_{Md} + p_{M1}\left[1 - \frac{\mu[p_{a1} + K_{a12}p_{a2}]}{1 + \mu p_{a1} + \mu p_{a2}}\right]^2$$

$$+ p_{M2}\left[1 - \frac{\mu[K_{a21}p_{a1} + p_{a2}]}{1 + \mu p_{a1} + \mu p_{a2}}\right]^2$$

Upon simplification, (C.7) gives

$$4p_{oSS} = p_{Md} + p_{M1}\frac{[1 + \mu p_{a2}(1 - K_{a12})]^2}{[1 + \mu p_{a1} + \mu p_{a2}]^2} + p_{M2}\frac{[1 + \mu p_{a1}(1 - K_{a21})]^2}{[1 + \mu p_{a1} + \mu p_{a2}]^2} \tag{C.8}$$

Appendix D

Final Output Power for a Three-Source, Three-Output System

From (2.111), the final steady-state output power p_{oSS}, appearing at the output port of the signal processing network, is given by

$$(2N)p_{oSS} = \left[\frac{mW_{aSS}\,y_a(t)}{\sqrt{2}} + \frac{mW_{bSS}\,y_b(t)}{\sqrt{2}} + y_M(t) \right]^2$$

$$= \frac{m^2 W_{aSS}^2 |y_a(t)|^2}{2} + \frac{m^2 W_{bSS}^2 |y_b(t)|^2}{2} + |y_M(t)|^2$$

$$+ \frac{2m^2 \mathrm{Re}[W_{aSS} W_{bSS}\, y_a(t)\, y_b(t)]}{2} \qquad (D.1)$$

$$+ \frac{2m\,\mathrm{Re}[W_{aSS}\, y_a(t)\, y_M(t)]}{\sqrt{2}}$$

$$+ \frac{2m\,\mathrm{Re}[W_{bSS}\, y_b(t)\, y_M(t)]}{\sqrt{2}}$$

Again, using (2.101),

$$|y_a(t)|^2 = |G_a(\theta_1)|^2 p_1 + |G_a(\theta_2)|^2 p_2$$

$$|y_b(t)|^2 = |G_b(\theta_1)|^2 p_1 + |G_b(\theta_2)|^2 p_2$$

$$|y_M(t)|^2 = |G_M(0)|^2 p_d + |G_M(\theta_1)|^2 p_1 + |G_M(\theta_2)|^2 p_2$$

$$|y_a(t)\,y_b(t)| = |G_a(\theta_1)\,G_b(\theta_1)|\,p_1 + |G_a(\theta_2)\,G_b(\theta_2)|\,p_2$$
$$|y_a(t)\,y_M(t)| = |G_a(\theta_1)\,G_M(\theta_1)|\,p_1 + |G_a(\theta_2)\,G_M(\theta_2)|\,p_2$$
$$|y_b(t)\,y_M(t)| = |G_b(\theta_1)\,G_M(\theta_1)|\,p_1 + |G_b(\theta_2)\,G_M(\theta_2)|\,p_2$$

where $p_d = |f_d(t)|^2$, $p_1 = |f_1(t)|^2$, and $p_2 = |f_2(t)|^2$.
Then, (D.1) becomes

$$
(2N)p_{oSS} = \frac{m^2 W_{aSS}^2[\,|G_a(\theta_1)|^2 p_1 + |G_a(\theta_2)|^2 p_2]}{2} \tag{D.2}
$$

$$
+ \frac{m^2 W_{bSS}^2[\,|G_b(\theta_1)|^2 p_1 + |G_b(\theta_2)|^2 p_2]}{2}
$$

$$
+ |G_M(0)|^2 p_d + |G_M(\theta_1)|^2 p_1 + |G_M(\theta_2)|^2 p_2
$$

$$
+ \frac{2m^2 W_{aSS} W_{bSS}[\,|G_a(\theta_1)\,G_b(\theta_1)|\,p_1 + |G_a(\theta_2)\,G_b(\theta_2)|\,p_2]}{2}
$$

$$
+ \frac{2m W_{aSS}[\,|G_a(\theta_1)\,G_M(\theta_1)|\,p_1 + |G_a(\theta_2)\,G_M(\theta_2)|\,p_2]}{\sqrt{2}}
$$

$$
+ \frac{2m W_{bSS}[\,|G_b(\theta_1)\,G_M(\theta_1)|\,p_1 + |G_b(\theta_2)\,G_M(\theta_2)|\,p_2]}{\sqrt{2}}
$$

Again, put

$$
K = 1 + \mu(p_{a1} + p_{a2}) + \mu(p_{b1} + p_{b2}) \tag{D.3}
$$

Then, (2.111) becomes

$$
W_{aSS} = -\frac{\mu(\sqrt{2})}{m} \frac{|G_a(\theta_1)\,G_M(\theta_1)|\,p_1 + |G_a(\theta_2)\,G_M(\theta_2)|\,p_2}{K} \tag{D.4a}
$$

$$
W_{bSS} = -\frac{\mu(\sqrt{2})}{m} \frac{|G_b(\theta_1)\,G_M(\theta_1)|\,p_1 + |G_b(\theta_2)\,G_M(\theta_2)|\,p_2}{K} \tag{D.4b}
$$

Substituting (D.4) into (D.2) and multiplying give

$$(2N)p_{oSS} = |G_M(0)|^2 p_d + |G_M(\theta_1)|^2 p_1 + |G_M(\theta_2)|^2 p_2$$

$$+ \mu^2 \frac{[|G_a(\theta_1)G_M(\theta_1)|p_1 + |G_a(\theta_2)G_M(\theta_2)|p_2]^2}{K^2} [|G_a(\theta_1)|^2 p_1]$$

$$+ \mu^2 \frac{[|G_a(\theta_1)G_M(\theta_1)|p_1 + |G_a(\theta_2)G_M(\theta_2)|p_2]^2}{K^2} [|G_a(\theta_2)|^2 p_2]$$

$$+ \mu^2 \frac{[|G_b(\theta_1)G_M(\theta_1)|p_1 + |G_b(\theta_2)G_M(\theta_2)|p_2]^2}{K^2} [|G_b(\theta_1)|^2 p_1]$$

$$+ \mu^2 \frac{[|G_b(\theta_1)G_M(\theta_1)|p_1 + |G_b(\theta_2)G_M(\theta_2)|p_2]^2}{K^2} [|G_b(\theta_2)|^2 p_2]$$

$$+ 2\mu^2 \frac{[|G_a(\theta_1)G_M(\theta_1)|p_1 + |G_a(\theta_2)G_M(\theta_2)|p_2]}{K}$$

$$\times \frac{[|G_b(\theta_1)G_M(\theta_1)|p_1 + |G_b(\theta_2)G_M(\theta_2)|p_2]}{K}$$

$$\times [|G_a(\theta_1)G_b(\theta_1)|p_1] \tag{D.5}$$

$$+ 2\mu^2 \frac{[|G_a(\theta_1)G_M(\theta_1)|p_1 + |G_a(\theta_2)G_M(\theta_2)|p_2]}{K}$$

$$\times \frac{[|G_b(\theta_1)G_M(\theta_1)|p_1 + |G_b(\theta_2)G_M(\theta_2)|p_2]}{K}$$

$$\times [|G_a(\theta_2)G_b(\theta_2)|p_2]$$

$$- 2\mu \frac{[|G_a(\theta_1)G_M(\theta_1)|p_1 + |G_a(\theta_2)G_M(\theta_2)|p_2]}{K} [|G_a(\theta_1)G_M(\theta_1)|p_1]$$

$$- 2\mu \frac{[|G_a(\theta_1)G_M(\theta_1)|p_1 + |G_a(\theta_2)G_M(\theta_2)|p_2]}{K} [|G_a(\theta_2)G_b(\theta_2)|p_2]$$

$$- 2\mu \frac{[|G_b(\theta_1)G_M(\theta_1)|p_1 + |G_b(\theta_2)G_M(\theta_2)|p_2]}{K} [|G_a(\theta_1)G_M(\theta_1)|p_1]$$

$$- 2\mu \frac{[|G_b(\theta_1)G_M(\theta_1)|p_1 + |G_b(\theta_2)G_M(\theta_2)|p_2]}{K} [|G_a(\theta_2)G_b(\theta_2)|p_2]$$

Rearranging the gain functions and simplifying, (D.5) reduces to

$$(2N)p_{oSS} = p_{Md}$$

$$+ p_{M1}\left[1 - \frac{\mu p_{a1} + \mu p_{a2}K_{a12}}{K} - \frac{\mu p_{b1} + \mu p_{b2}K_{b12}}{K}\right]^2 \quad (D.6)$$

$$+ p_{M2}\left[1 - \frac{\mu p_{a1}K_{a21} + \mu p_{a2}}{K} - \frac{\mu p_{b1}K_{b21} + \mu p_{b2}}{K}\right]^2$$

Substituting (D.3) into (D.6) and simplifying gives

$$(2N)p_{oSS} = p_{Md} + p_{M1}\frac{[1 + \mu p_{a2}(1 - K_{a12}) + \mu p_{b2}(1 - K_{b12})]^2}{[1 + \mu(p_{a1} + p_{b1} + p_{a2} + p_{b2})]^2} \quad (D.7)$$

$$+ p_{M2}\frac{[1 + \mu p_{a1}(1 - K_{a21}) + \mu p_{b1}(1 - K_{b21})]^2}{[1 + \mu(p_{a1} + p_{b1} + p_{a2} + p_{b2})]^2}$$

Equation (D.7) is the same as (2.115a).

Appendix E

Final Output Power for a Multisource, Two-Output System

From (2.132), the final output power p_{oSS}, appearing at the output port of the signal processing network, is given by

$$4p_{oSS} = \frac{m^2 W_{ASS}^2 |y_A(t)|^2}{2} + |y_M(t)|^2 + \frac{2m\,\mathrm{Re}[W_{ASS}\,y_A(t)\,y_M(t)]}{\sqrt{2}}$$

(E.1)

Using (2.121),

$$|y_A(t)|^2 = |G_A(\theta_1)|^2 p_1 + |G_A(\theta_2)|^2 p_2 + |G_A(\theta_3)|^2 p_3 + \ldots$$

$$|y_M(t)|^2 = |G_M(\theta_1)|^2 p_1 + |G_M(\theta_2)|^2 p_2 + |G_M(\theta_3)|^2 p_3 + \ldots$$

$$|y_A(t)\,y_M(t)| = |G_A(\theta_1)G_M(\theta_1)|\,p_1 + |G_A(\theta_2)G_M(\theta_2)|\,p_2$$
$$+ |G_A(\theta_3)G_M(\theta_3)|\,p_3 + \ldots$$

where

$$p_1 = |f_1(t)|^2 = \text{power of interference signal 1;}$$

$$p_2 = |f_2(t)|^2 = \text{power of interference signal 2;}$$

$$p_3 = |f_3(t)|^2 = \text{power of interference signal 3;}$$

$$\ldots \qquad \ldots \qquad \ldots$$

Then, (E.1) becomes

$$4p_{oSS} = \frac{m^2 W_{ASS}^2[\,|G_A(\theta_1)|^2 p_1 + |G_A(\theta_2)|^2 p_2 + |G_A(\theta_3)|^2 p_3 + \ldots\,]}{2}$$

$$+ [\,|G_M(\theta_1)|^2 p_1 + |G_M(\theta_2)|^2 p_2 + |G_M(\theta_3)|^2 p_3 + \ldots\,] \qquad \text{(E.2)}$$

$$+ \frac{2m W_{ASS}[\,|G_A(\theta_1)G_M(\theta_1)|\,p_1 + |G_A(\theta_2)G_M(\theta_2)|\,p_2 + |G_A(\theta_3)G_M(\theta_3)|\,p_3 + \ldots\,]}{\sqrt{2}}$$

Again, from (2.131), the steady-state weighter is given by

$$W_{ASS} = -\frac{(\sqrt{2})[\,|G_A(\theta_1)G_M(\theta_1)|\,p_1 + |G_A(\theta_2)G_M(\theta_2)|\,p_2 + \ldots\,]}{m[(1/\mu) + p_{A1} + p_{A2} + p_{A3} + \ldots\,]}$$

$$\text{(E.3)}$$

Substitution of (E.3) into (E.2) gives

$$4p_{oSS} = \frac{[\,|G_A(\theta_1)G_M(\theta_1)|\,p_1 + |G_A(\theta_2)G_M(\theta_2)|\,p_2 + \ldots\,]^2}{[(1/\mu) + p_{A1} + p_{A2} + p_{A3} + \ldots\,]^2} \cdot [\,|G_A(\theta_1)|^2 p_1 + |G_A(\theta_2)|^2 p_2 + \ldots\,] \qquad \text{(E.4)}$$

$$+ [\,|G_M(\theta_1)|^2 p_1 + |G_M(\theta_2)|^2 p_2 + |G_M(\theta_3)|^2 p_3 + \ldots\,]$$

$$- \frac{2[\,|G_A(\theta_1)G_M(\theta_1)|\,p_1 + |G_A(\theta_2)G_M(\theta_2)|\,p_2 + \ldots\,] \, [\,|G_A(\theta_1)G_M(\theta_1)|\,p_1 + |G_A(\theta_2)G_M(\theta_2)|\,p_2 + \ldots\,]}{[(1/\mu) + p_{A1} + p_{A2} + p_{A3} + \ldots\,]}$$

$$= [\,|G_M(\theta_1)|^2 p_1 + |G_M(\theta_2)|^2 p_2 + |G_M(\theta_3)|^2 p_3 + \ldots\,]$$

$$+ \frac{[\,|G_A(\theta_1)G_M(\theta_1)|\,p_1 + |G_A(\theta_2)G_M(\theta_2)|\,p_2 + |G_A(\theta_3)G_M(\theta_3)|\,p_3 + \ldots\,]^2}{[(1/\mu) + p_{A1} + p_{A2} + p_{A3} + \ldots\,]^2}\,|G_A(\theta_1)|^2 p_1$$

$$+ \frac{[\,|G_A(\theta_1)G_M(\theta_1)|\,p_1 + |G_A(\theta_2)G_M(\theta_2)|\,p_2 + |G_A(\theta_3)G_M(\theta_3)|\,p_3 + \ldots\,]^2}{[(1/\mu) + p_{A1} + p_{A2} + p_{A3} + \ldots\,]^2}\,|G_A(\theta_2)|^2 p_2$$

$$+\frac{[\,|\,G_A(\theta_1)G_M(\theta_1)\,|\,p_1 + |\,G_A(\theta_2)G_M(\theta_2)\,|\,p_2 + |\,G_A(\theta_3)G_M(\theta_3)\,|\,p_3 + \ldots\,]^2}{[\,(1/\mu) + p_{A1} + p_{A2} + p_{A3} + \ldots\,]^2}\,|\,G_A(\theta_3)\,|^2 p_3$$

$$+\ldots \qquad \ldots \qquad \ldots$$

$$-\frac{2[\,|\,G_A(\theta_1)G_M(\theta_1)\,|\,p_1 + |\,G_A(\theta_2)G_M(\theta_2)\,|\,p_2 + |\,G_A(\theta_3)G_M(\theta_3)\,|\,p_3 + \ldots\,]}{[\,(1/\mu) + p_{A1} + p_{A2} + p_{A3} + \ldots\,]^2}\,|\,G_A(\theta_1)G_M(\theta_1)\,|\,p_1$$

$$-\frac{2[\,|\,G_A(\theta_1)G_M(\theta_1)\,|\,p_1 + |\,G_A(\theta_2)G_M(\theta_2)\,|\,p_2 + |\,G_A(\theta_3)G_M(\theta_3)\,|\,p_3 + \ldots\,]}{[\,(1/\mu) + p_{A1} + p_{A2} + p_{A3} + \ldots\,]^2}\,|\,G_A(\theta_2)G_M(\theta_1)\,|\,p_2$$

$$-\frac{2[\,|\,G_A(\theta_1)G_M(\theta_1)\,|\,p_1 + |\,G_A(\theta_2)G_M(\theta_2)\,|\,p_2 + |\,G_A(\theta_3)G_M(\theta_3)\,|\,p_3 + \ldots\,]}{[\,(1/\mu) + p_{A1} + p_{A2} + p_{A3} + \ldots\,]^2}\,|\,G_A(\theta_3)G_M(\theta_1)\,|\,p_3$$

$$-\ldots \qquad \ldots \qquad \ldots$$

Rearranging the gain functions (E.4) gives

$$4p_{oSS} = [\,|\,G_M(\theta_1)\,|^2 p_1 + |\,G_M(\theta_2)\,|^2 p_2 + |\,G_M(\theta_3)\,|^2 p_3 + \ldots\,] \qquad (E.5)$$

$$+\frac{[\,|\,G_A(\theta_1)G_M(\theta_1)\,|\,p_1 + |\,G_A(\theta_2)G_M(\theta_2)\,|\,p_2 + |\,G_A(\theta_3)G_M(\theta_3)\,|\,p_3 + \ldots\,]^2}{[\,(1/\mu) + p_{A1} + p_{A2} + p_{A3} + \ldots\,]^2}$$

$$\times [\,|\,G_A(\theta_1)\,|^2 p_1 + |\,G_A(\theta_2)\,|^2 p_2 + |\,G_A(\theta_3)\,|^2 p_3 + \ldots\,]$$

$$-\frac{2[\,|\,G_A(\theta_1)G_M(\theta_1)\,|\,p_1 + |\,G_A(\theta_2)G_M(\theta_2)\,|\,p_2 + |\,G_A(\theta_3)G_M(\theta_3)\,|\,p_3 + \ldots\,]}{[\,(1/\mu) + p_{A1} + p_{A2} + p_{A3} + \ldots\,]^2}$$

$$\times [\,|\,G_A(\theta_1)G_M(\theta_1)\,|\,p_1 + |\,G_A(\theta_2)G_M(\theta_2)\,|\,p_2 + |\,G_A(\theta_3)G_M(\theta_3)\,|\,p_3 + \ldots\,]$$

$$= [\,|\,G_M(\theta_1)\,|^2 p_1 + |\,G_M(\theta_2)\,|^2 p_2 + |\,G_M(\theta_3)\,|^2 p_3 + \ldots\,]$$

$$+\frac{[\,|\,G_A(\theta_1)G_M(\theta_1)\,|\,p_1 + |\,G_A(\theta_2)G_M(\theta_2)\,|\,p_2 + |\,G_A(\theta_3)G_M(\theta_3)\,|\,p_3 + \ldots\,]^2}{[\,(1/\mu) + p_{A1} + p_{A2} + p_{A3} + \ldots\,]^2}\,|\,G_A(\theta_1)\,|^2 p_1$$

$$+ \frac{[|G_A(\theta_1)G_M(\theta_1)|p_1 + |G_A(\theta_2)G_M(\theta_2)|p_2 + |G_A(\theta_3)G_M(\theta_3)|p_3 + \ldots]^2}{[(1/\mu) + p_{A1} + p_{A2} + p_{A3} + \ldots]^2} |G_A(\theta_2)|^2 p_2$$

$$+ \frac{[|G_A(\theta_1)G_M(\theta_1)|p_1 + |G_A(\theta_2)G_M(\theta_2)|p_2 + |G_A(\theta_3)G_M(\theta_3)|p_3 + \ldots]^2}{[(1/\mu) + p_{A1} + p_{A2} + p_{A3} + \ldots]^2} |G_A(\theta_3)|^2 p_3$$

$$+ \ldots \qquad \ldots \qquad \ldots$$

$$- \frac{2[|G_A(\theta_1)G_M(\theta_1)|p_1 + |G_A(\theta_2)G_M(\theta_2)|p_2 + |G_A(\theta_3)G_M(\theta_3)|p_3 + \ldots]}{[(1/\mu) + p_{A1} + p_{A2} + p_{A3} + \ldots]^2}$$
$$\times |G_A(\theta_1)G_M(\theta_1)|p_1$$

$$- \frac{2[|G_A(\theta_1)G_M(\theta_1)|p_1 + |G_A(\theta_2)G_M(\theta_2)|p_2 + |G_A(\theta_3)G_M(\theta_3)|p_3 + \ldots]}{[(1/\mu) + p_{A1} + p_{A2} + p_{A3} + \ldots]^2}$$
$$\times |G_A(\theta_2)G_M(\theta_2)|p_2$$

$$- \frac{2[|G_A(\theta_1)G_M(\theta_1)|p_1 + |G_A(\theta_2)G_M(\theta_2)|p_2 + |G_A(\theta_3)G_M(\theta_3)|p_3 + \ldots]}{[(1/\mu) + p_{A1} + p_{A2} + p_{A3} + \ldots]^2}$$
$$\times |G_A(\theta_3)G_M(\theta_3)|p_3$$

$$- \ldots \qquad \ldots \qquad \ldots$$

Rearranging the gain functions, (E.5) gives

$$4p_{oSS} = p_{Md} + p_{M1} + p_{M2} + p_{M3} + \ldots$$
$$+ p_{M1} \frac{[p_{A1} + K_{12}p_{A2} + K_{13}p_{A3} + \ldots]^2}{[(1/\mu) + p_{A1} + p_{A2} + p_{A3} + \ldots]^2}$$
$$+ p_{M2} \frac{[K_{21}p_{A1} + p_{A2} + K_{23}p_{A3} + \ldots]^2}{[(1/\mu) + p_{A1} + p_{A2} + p_{A3} + \ldots]^2}$$
$$+ p_{M3} \frac{[K_{31}p_{A1} + K_{32}p_{A2} + p_{A3} + \ldots]^2}{[(1/\mu) + p_{A1} + p_{A2} + p_{A3} + \ldots]^2}$$
$$+ \ldots \qquad \ldots \qquad \ldots \qquad \text{(E.6)}$$

$$- 2p_{M1} \frac{[p_{A1} + K_{12}p_{A2} + K_{13}p_{A3} + \ldots]}{[(1/\mu) + p_{A1} + p_{A2} + p_{A3} + \ldots]}$$

$$- 2p_{M2} \frac{[K_{21}p_{A1} + p_{A2} + K_{23}p_{A3} + \ldots]}{[(1/\mu) + p_{A1} + p_{A2} + p_{A3} + \ldots]}$$

$$- 2p_{M3} \frac{[K_{31}p_{A1} + K_{32}p_{A2} + p_{A3} + \ldots]}{[(1/\mu) + p_{A1} + p_{A2} + p_{A3} + \ldots]}$$

$$- \ldots \qquad \ldots \qquad \ldots$$

where

$p_{Md} = |G_M(0)|^2 p_d$ = power of the desired signal at the main output port;

$p_{M1} = |G_M(\theta_1)|^2 p_1$ = power of the interference signal 1 at the main output port;

$p_{M2} = |G_M(\theta_2)|^2 p_2$ = power of the interference signal 2 at the main output port;

$p_{M3} = |G_M(\theta_3)|^2 p_1$ = power of the interference signal 3 at the main output port;

$$\ldots \qquad \ldots \qquad \ldots$$

$p_{A1} = |G_A(\theta_1)|^2 p_1$ = power of the interference signal 1 at the AUX-A output port;

$p_{A2} = |G_A(\theta_2)|^2 p_1$ = power of the interference signal 2 at the AUX-A output port;

$p_{A3} = |G_A(\theta_3)|^2 p_1$ = power of the interference signal 3 at the AUX-A output port.

$$\ldots \qquad \ldots \qquad \ldots$$

and where

$$K_{Amn} = \frac{1}{K_{Anm}} = \frac{|G_A(\theta_m)/|G_A(\theta_n)|}{|G_M(\theta_m)/|G_M(\theta_n)|}$$

Hence, the final steady-state output power in (E.6) can be written as

$$
\begin{aligned}
4p_{oSS} = p_{Md} + p_{M1} & \left[1 - \frac{[p_{A1} + K_{12}p_{A2} + K_{13}p_{A3} + \ldots]}{[(1/\mu) + p_{A1} + p_{A2} + p_{A3} + \ldots]} \right]^2 \\
+ p_{M2} & \left[1 - \frac{[K_{21}p_{A1} + p_{A2} + K_{23}p_{A3} + \ldots]}{[(1/\mu) + p_{A1} + p_{A2} + p_{A3} + \ldots]} \right]^2 \qquad \text{(E.7)} \\
+ p_{M3} & \left[1 - \frac{[K_{31}p_{A1} + K_{32}p_{A2} + p_{A3} + \ldots]}{[(1/\mu) + p_{A1} + p_{A2} + p_{A3} + \ldots]} \right]^2 \\
\ldots \qquad & \ldots \qquad \qquad \ldots
\end{aligned}
$$

Equation (E.7) is the same as (2.133).

Appendix F

Final Output Power in the Presence of Leakage

Substitution of (2.156) into (2.157) gives the final steady-state output power p_{oSS}, appearing at the output port of the signal processing network, in the presence of desired signal in the AUX output ports. Hence,

$$4p_{oSS} = [|G_M(0)|^2 p_d + |G_M(\theta_i)|^2 p_i] \tag{F.1}$$

$$+ \frac{[\mu|G_A(0)G_M(0)|p_d + \mu|G_A(\theta_i)G_M(\theta_i)|p_i]^2}{[1 + \mu|G_A(0)|^2 p_d + \mu|G_A(\theta_i)|^2 p_i]^2}$$

$$\times [|G_A(0)|^2 p_d + |G_A(\theta_i)|^2 p_i]$$

$$- \frac{2[\mu|G_A(0)G_M(0)|p_d + \mu|G_A(\theta_i)G_M(\theta_i)|p_i]}{[1 + \mu|G_A(0)|^2 p_d + \mu|G_A(\theta_i)|^2 p_i]}$$

$$\times [|G_A(0)G_M(0)|p_d + \mu|G_A(\theta_i)G_M(\theta_i)|p_i]$$

$$= p_{Md} + p_{Mi}$$

$$+ \frac{[m|G_A(0)|^2 p_d + mK_{A01}|G_A(\theta_i)|^2 p_i]^2}{[1 + \mu p_{Ad} + \mu p_{Ai}]^2} [|G_A(0)|^2 p_d]$$

$$+ \frac{[mK_{A10}|G_A(0)|^2 p_d + m|G_A(\theta_i)|^2 p_i]^2}{[1 + \mu p_{Ad} + \mu p_{Ai}]^2} [|G_M(\theta_i)|^2 p_i]$$

$$- \frac{2[m|G_A(0)|^2 p_d + mK_{A01}|G_A(\theta_i)|^2 p_i]^2}{[1 + \mu p_{Ad} + \mu p_{Ai}]^2} [|G_A(0)|^2 p_d]$$

$$- \frac{2[mK_{A10}|G_A(0)|^2 p_d + m|G_A(\theta_i)|^2 p_i]^2}{[1 + \mu p_{Ad} + \mu p_{Ai}]^2} [|G_M(\theta_i)|^2 p_i]$$

$$= p_{Md} + M_{Mi}$$

$$+ \frac{[mp_{Ad} + mK_{A01} p_{Ai}]^2}{[1 + \mu p_{Ad} + \mu p_{Ai}]^2} [p_{Md}]$$

$$+ \frac{[mK_{A10} p_{Ad} + mp_{Ai}]^2}{[1 + \mu p_{Ad} + \mu p_{Ai}]^2} [p_{Mi}]$$

$$- \frac{2[mp_{Ad} + mK_{A01} p_{Ai}]}{[1 + \mu p_{Ad} + \mu p_{Ai}]} [p_{Md}]$$

$$- \frac{2[mK_{A10} p_{Ad} + mp_{Ai}]}{[1 + \mu p_{Ad} + \mu p_{Ai}]} [p_{Mi}]$$

$$\text{where} \quad K_{A01} = \frac{1}{K_{A10}} = \frac{|G_A(0)/G_A(\theta_i)|}{|G_M(0)/G_M(\theta_i)|} \tag{F.2}$$

and

$p_{Md} = |G_M(0)|^2 p_d$ = desired signal power at the main output port;

$p_{Mi} = |G_M(\theta_i)|^2 p_i$ = interference signal power at the main output port;

$p_{Ad} = |G_A(0)|^2 p_d$ = desired signal power at the AUX-A output port;

$p_{Ai} = |G_A(\theta_i)|^2 p_i$ = interference signal power at the AUX-A output port.

Rearranging, (F.1) gives

$$4p_{oSS} = p_{Md} + \frac{[p_{Ad} + K_{A01} p_{Ai}]^2}{[1 + \mu p_{Ad} + \mu p_{Ai}]^2} [p_{Md}] - 2\frac{[p_{Ad} + K_{A01} p_{Ai}]}{[1 + \mu p_{Ad} + \mu p_{Ai}]} [p_{Md}]$$

$$+ p_{Mi} + \frac{[p_{Ai} + K_{A10} p_{Ad}]^2}{[1 + \mu p_{Ad} + \mu p_{Ai}]^2} [p_{Mi}] - 2\frac{[p_{Ai} + K_{A10} p_{Ad}]}{[1 + \mu p_{Ad} + \mu p_{Ai}]} [p_{Mi}]$$

$$= p_{Md}\left[1 - \frac{p_{Ad} + K_{A01} p_{Ai}}{1 + \mu p_{Ad} + \mu p_{Ai}}\right]^2 + p_{Mi}\left[1 - \frac{p_{Ai} + K_{A10} p_{Ad}}{1 + \mu p_{Ad} + \mu p_{Ai}}\right]^2 \tag{F.3}$$

Equation (F.3) is the same as (2.158a).

Appendix G

Final Output Power in the Presence of Offset Voltage

Replacing W_{AOFF} in (2.178) with the steady-state value W_{ASSO}, the final steady-state output power p_{oSSO}, appearing at the output port of the signal processing network in the presence of an offset voltage, is given by

$$4p_{oSSO} = 4p_o = \frac{m^2 W_{ASSO}^2 |G_A(\theta_i)|^2 p_i}{2} + |G_M(0)|^2 p_d + |G_M(\theta_i)|^2 p_i$$

$$+ \frac{2m W_{ASSO} |G_A(\theta_i) G_M(\theta_i)| p_i}{\sqrt{2}} \tag{G.1}$$

Substituting W_{ASSO} from (2.175) into (G.1) and rearranging gives

$$4p_{oSSO} = |G_M(0)|^2 p_d + |G_M(\theta_i)|^2 p_i$$

$$+ \frac{m^2}{2} \left[V_{OFF} + \frac{\mu\sqrt{2}|G_A(\theta_i) G_M(\theta_i)| p_i}{m[1 + \mu |G_A(\theta_i)|^2 p_i]} \right]^2 |G_A(\theta_i)|^2 p_i$$

$$- \frac{2m}{\sqrt{2}} \left[V_{OFF} + \frac{\mu(\sqrt{2})|G_A(\theta_i) G_M(\theta_i)| p_i}{m[1 + \mu |G_A(\theta_i)|^2 p_i]} \right] |G_A(\theta_i) G_M(\theta_i)| p_i$$

$$= |G_M(0)|^2 p_d + |G_M(\theta_i)|^2 p_i$$

$$+ \frac{m^2}{2} \left[V_{OFF} \frac{G_A(\theta_i)}{G_M(\theta_i)} + \frac{\mu\sqrt{2}|G_A(\theta_i)|^2 p_i}{m[1 + \mu |G_A(\theta_i)|^2 p_i]} \right]^2 |G_M(\theta_i)|^2 p_i$$

$$-\frac{2m}{\sqrt{2}}\left[V_{OFF}\frac{G_A(\theta_i)}{G_M(\theta_i)}+\frac{\mu\sqrt{2}|G_A(\theta_i)|^2 p_i}{m[1+\mu|G_A(\theta_i)|^2 p_i]}\right]|G_M(\theta_i)|^2 p_i$$

$$=p_{Md}+p_{Mi}+\frac{m^2}{2}\left[\frac{V_{OFF}(\sqrt{p_{Ai}})}{(\sqrt{p_{Mi}})}+\frac{\mu\sqrt{2}p_{Ai}}{m(1+\mu p_{Ai})}\right]^2 p_{Mi}$$

$$-\frac{2m}{\sqrt{2}}\left[\frac{V_{OFF}(\sqrt{p_{Ai}})}{(\sqrt{p_{Mi}})}+\frac{\mu(\sqrt{2})p_{Ai}}{m(1+\mu p_{Ai})}\right]p_{Mi}$$

$$=p_{Md}+p_{Mi}\left[1-\frac{m}{\sqrt{2}}\left[\frac{V_{OFF}(\sqrt{p_{Ai}})}{(\sqrt{p_{Mi}})}+\frac{\mu(\sqrt{2})p_{Ai}}{m(1+\mu p_{Ai})}\right]\right]^2$$

$$=p_{Md}+p_{Mi}\left[1-\frac{\mu p_{Ai}}{(1+\mu p_{Ai})}+\frac{m}{\sqrt{2}}\frac{V_{OFF}(\sqrt{p_{Ai}})}{(\sqrt{p_{Mi}})}\right]^2$$

$$=p_{Md}+p_{Mi}\left[\frac{1}{(1+\mu p_{Ai})}+\frac{m}{\sqrt{2}}\frac{V_{OFF}(\sqrt{p_{Ai}})}{(\sqrt{p_{Mi}})}\right]^2$$

$$=p_{Md}+\frac{p_{Mi}}{(1+\mu p_{Ai})}\left[1+(1+\mu p_{Ai})\frac{m}{\sqrt{2}}\frac{V_{OFF}(\sqrt{p_{Ai}})}{(\sqrt{p_{Mi}})}\right]^2$$

$$(G.2)$$

Appendix **H**

Final Output Power in the Presence of Space Noise

The final steady state output power p_{oSSn}, appearing at the output port of the signal processing network in the presence of space noise, can be obtained by replacing the weighter W_A in (2.197) by W_{ASS} from (2.155) to give

$$4p_{oSSO} = |G_M(0)|^2 p_d + |G_M(\theta_i)|^2 p_i + |G_M(0)|^2 p_n \qquad \text{(H.1)}$$

$$+ \frac{[\mu|G_A(\theta_i)G_M(\theta_i)|\,p_i + \mu|G_A(\theta_A)G_M(0)|\,p_n]^2}{[1 + \mu|G_A(\theta_i)|^2 p_i + \mu|G_A(\theta_A)|^2 p_n]^2}$$

$$\times [|G_A(\theta_i)|^2 p_i + |G_A(\theta_A)|^2 p_n]$$

$$- \frac{2[\mu|G_A(\theta_i)G_M(\theta_i)|\,p_i + \mu|G_A(\theta_A)G_M(0)|\,p_n]}{[1 + \mu|G_A(\theta_i)|^2 p_i + \mu|G_A(\theta_A)|^2 p_n]}$$

$$\times [\mu|G_A(\theta_i)G_M(\theta_i)|\,p_i + \mu|G_A(\theta_A)G_M(0)|\,p_n]$$

$$= |G_M(0)|^2 p_d + |G_M(\theta_i)|^2 p_i + |G_M(0)|^2 p_n$$

$$+ \frac{\left\{\mu|G_A(\theta_i)|^2 p_i + \mu\left[\dfrac{|G_A(\theta_i)/G_A(\theta_A)|}{|G_M(\theta_i)/G_M(0)|}\right]|G_A(\theta_A)|^2 p_n\right\}^2}{[1 + \mu p_{Ai} + \mu p_{An}]^2} |G_M(\theta_i)|^2 p_i$$

$$+ \frac{\left\{\mu\left[\dfrac{|G_A(\theta_A)/G_A(\theta_i)|}{|G_M(0)/G_M(\theta_i)|}\right]|G_A(\theta_i)|^2 p_i + |G_A(\theta_A)|^2 p_n\right\}^2}{[1 + \mu p_{Ai} + \mu p_{An}]^2} |G_M(0)|^2 p_n$$

$$- \frac{2\left\{ \mu |G_A(\theta_i)|^2 p_i + \mu \left[\frac{|G_A(\theta_i)/G_A(\theta_A)|}{|G_M(\theta_i)/G_M(0)|} \right] |G_A(\theta_A)|^2 p_n \right\}}{[1 + \mu p_{Ai} + \mu p_{An}]} |G_M(\theta_i)|^2 p_i$$

$$- \frac{2\left\{ \mu \left[\frac{|G_A(\theta_A)/G_A(\theta_i)|}{|G_M(0)/G_M(\theta_i)|} \right] |G_A(\theta_i)|^2 p_i + |G_A(\theta_A)|^2 p_n \right\}}{[1 + \mu p_{Ai} + \mu p_{An}]} |G_M(0)|^2 p_n$$

$$= p_{Md} + p_{Mi} + p_{Mn}$$

$$+ \frac{[\mu p_{Ai} + \mu p_{An} K_{AIA}]^2}{[1 + \mu p_{Ai} + \mu p_{An}]^2} p_{Mi} + \frac{[\mu p_{Ai} K_{AAI} + \mu p_{An}]^2}{[1 + \mu p_{Ai} + \mu p_{An}]^2} p_{Mn}$$

$$- \frac{2[\mu p_{Ai} + \mu p_{An} K_{AIA}]}{[1 + \mu p_{Ai} + \mu p_{An}]} p_{Mi} - \frac{2[\mu p_{Ai} K_{AAI} + \mu p_{An}]}{[1 + \mu p_{Ai} + \mu p_{An}]} p_{Mn}$$

$$= p_{Md} + p_{Mi} \left[1 + \frac{\mu p_{Ai} + \mu p_{An} K_{AIA}}{1 + \mu p_{Ai} + \mu p_{An}} \right]^2 + p_{Mn} \left[1 + \frac{\mu p_{Ai} K_{AAI} + \mu p_{An}}{1 + \mu p_{Ai} + \mu p_{An}} \right]^2$$

where

$p_{Md} = |G_M(0)|^2 p_d$ = the desired signal power at the main output port;

$p_{Mi} = |G_M(\theta_i)|^2 p_i$ = the interference signal power at the main output port;

$p_{Mn} = |G_M(0)|^2 p_n$ = the average noise power at the main output port;

$p_{Ai} = |G_A(\theta_i)|^2 p_i$ = the interference signal power at the AUX-A output port;

$p_{An} = |G_A(\theta_A)|^2 p_n$ = the average noise power at the AUX-A output port;

and $K_{AIA} = \dfrac{1}{K_{AAI}} = \dfrac{|G_A(\theta_i)|/|G_A(\theta_A)|}{|G_M(\theta_i)|/|G_M(0)|} = \dfrac{|G_A(\theta_i)|}{|G_M(\theta_i)|}$

where the normalized gain along the peak of a beam is unity. Equation (H.1) is the same as (2.198a).

Appendix I

Basic Matrix Relations

Some basic matrix relations used in this book are listed here. They have been taken from the book *Standard Mathematical Tables,* published by Chemical Rubber Company [1].

1. A matrix of order $m \times n$ has m rows and n columns and is called a *rectangular matrix.* In special cases where $m = n$, it is called a *square matrix.*

2. The transpose of a matrix S, denoted by S^T, is obtained by interchanging the rows and columns of matrix S:

$$S = \begin{bmatrix} S_{11} & S_{12} & S_{13} \\ S_{21} & S_{22} & S_{23} \\ S_{31} & S_{32} & S_{33} \end{bmatrix} \quad S^T = \begin{bmatrix} S_{11} & S_{21} & S_{31} \\ S_{12} & S_{22} & S_{32} \\ S_{13} & S_{23} & S_{33} \end{bmatrix}$$

3. The Hermitian conjugate of a matrix S, denoted by S^H, is obtained by replacing each element in the transpose of S by its complex conjugate. Thus,

$$S^H = S^{T}*$$

4. A square matrix is called symmetric if $S = S^T$ and Hermitian if $S = S^H$.

5. A matrix with m rows and one column ($m \times 1$) is called a *column matrix* or *vector.* So, if S_1 is a column matrix of m rows, then

$$S_1 = \begin{bmatrix} S_{11} \\ S_{21} \\ S_{m1} \end{bmatrix}$$

6. Two matrices A and B can be multiplied if the number of columns in A is equal to the number of rows in B. Thus, if $A = m \times n$ and $B = n \times p$, then $AB = m \times p$.

7. A square matrix with all off-diagonal elements equal to zero is called a *diagonal matrix*.

$$D_S = \begin{bmatrix} S_{11} & 0 & 0 \\ 0 & S_{22} & 0 \\ 0 & 0 & S_{33} \end{bmatrix}$$

8. A diagonal matrix with all the elements equal to one is called a unit matrix I.

$$I = \begin{vmatrix} 1 & 0 & 0 \\ 0 & 1 & 0 \\ 0 & 0 & 1 \end{vmatrix}$$

9. If M is a square matrix, then $|M - \lambda I| = 0$ is the characteristic equation of the matrix M.

10. If the order of matrix M is $n \times n$, then the characteristic equation has n roots called the *characteristic roots* or *eigenvalues* of M and are given by $\lambda_1, \lambda_2, \lambda_3, \ldots, \lambda_n$.

11. If λ_n is one of the characteristic roots of a matrix M, then the equation $Me_n = \lambda_n e_n$ will have a nonzero solution, with e_n called a *characteristic vector*. If $e_n^T e_n = 1$ or $e_n^H e_n = 1$, then e_n is called an eigenvector associated with the eigenvalue λ_n of matrix M.

12. The matrix $E = |e_1 \ e_2 \ \ldots \ e_n|$ of eigenvectors e_n of a real symmetric matrix is orthogonal.

13. If Λ is the diagonal matrix of the eigenvalue λ, then $ME = E\Lambda$.

14. For a real symmetric matrix, $M = E\Lambda E^H$ (matrix decomposition).

15. If M is symmetric, then $E^H = E^{-1}$ and so $M = E\Lambda E^{-1} = E\Lambda E^H$.

16. If A is a 3×3 matrix and B is a 3×3 matrix, then the product AB is a 3×3 matrix:

$$A = \begin{bmatrix} a_{11} & a_{12} & a_{13} \\ a_{21} & a_{22} & a_{23} \\ a_{31} & a_{32} & a_{33} \end{bmatrix} \quad B = \begin{bmatrix} b_{11} & b_{12} & b_{13} \\ b_{21} & b_{22} & b_{23} \\ b_{31} & b_{32} & b_{33} \end{bmatrix}$$

$AB =$

$$\begin{bmatrix} a_{11}b_{11} + a_{12}b_{21} + a_{13}b_{31} & a_{11}b_{12} + a_{12}b_{22} + a_{13}b_{23} & a_{11}b_{13} + a_{12}b_{23} + a_{13}b_{33} \\ a_{21}b_{11} + a_{22}b_{21} + a_{23}b_{31} & a_{21}b_{12} + a_{22}b_{22} + a_{23}b_{32} & a_{21}b_{13} + a_{22}b_{23} + a_{23}b_{33} \\ a_{31}b_{11} + a_{32}b_{21} + a_{33}b_{31} & a_{31}b_{12} + a_{32}b_{22} + a_{33}b_{32} & a_{31}b_{13} + a_{32}b_{23} + a_{33}b_{33} \end{bmatrix}$$

17. If A is a 2 × 3 matrix (3 columns) and B is a 3 × 3 matrix (3 rows), then the product AB is a 2 × 3 matrix.

$$A = \begin{bmatrix} a_{11} & a_{12} & a_{13} \\ a_{21} & a_{22} & a_{23} \end{bmatrix} \quad B = \begin{bmatrix} b_{11} & b_{12} & b_{13} \\ b_{21} & b_{22} & b_{23} \\ b_{31} & b_{32} & b_{33} \end{bmatrix}$$

$AB =$

$$\begin{bmatrix} a_{11}b_{11} + a_{12}b_{21} + a_{13}b_{31} & a_{11}b_{12} + a_{12}b_{22} + a_{13}b_{23} & a_{11}b_{13} + a_{12}b_{23} + a_{13}b_{33} \\ a_{21}b_{11} + a_{22}b_{21} + a_{23}b_{31} & a_{21}b_{12} + a_{22}b_{22} + a_{23}b_{32} & a_{21}b_{13} + a_{22}b_{23} + a_{23}b_{33} \end{bmatrix}$$

Reference

[1] Selby, S. M. (ed.), *Standard Mathematical Tables*, Chemical Rubber Company: Cleveland, OH, 1969, pp. 118–140.

Appendix J

Touchstone-Generated Phase Values for Cross-Coupled Transmission Lines

As mentioned in Section 2.14.2.2, the expression for the phase α at the junction of a pair of cross-coupled transmission lines becomes increasingly inaccurate at frequencies far away from the center frequency. Under the circumstances, to obtain a more reliable result, the physical structure of the cross-coupled transmission lines has been analyzed by using Touchstone software.

The physical structure under consideration is shown in Figure J.1 and a typical Touchstone program is shown in Figure J.2.

In Tables J.1 through J.6, ANG(S_{21}) represents the phase ϕ_C at the output of the cross-coupled lines as given by (2.285). In addition, the insertion loss S_{21} and the return loss S_{11} have also been calculated.

Tables J.1 and J.2 show the phase values for $\theta = 30°$; those in Tables J.3 and J.4 are for $\theta = 45°$, while the results in Tables J.5 and J.6 are for $\theta = 60°$.

In Tables J.1, J.3, and J.5, both Z_1 and Z_0 have the same value. In Tables J.2, J.4, and J.6, the value of Z_1 has been kept constant at 32Ω while that of Z_0 has been varied.

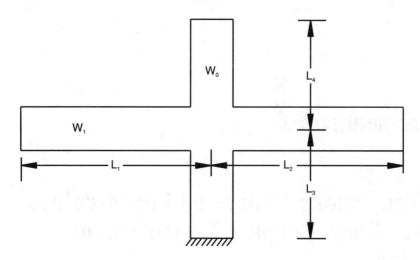

Figure J.1 Cross-coupled line parameters used in Tables J.1 through J.6.

```
VAR
!Z0=25/Z1=25/L1,L2=90/L3=30/L4=60
            W0=482
            W1=482
            L1= 835
            L2=835
            L3=348
            L4=610

DIM
        FREQ        GHZ
        LNG         MIL

CKT
        MSUB    ER=2.2   H=62   T=67   RHO=1   TAND=2.8E-3

!PS
        MLIN        1    21       W^W1     L^L1
        MCROS       21   22       23       24      W1^W1   W2^W0   W3^W1   W4^W3
        MLOC        22   W^W0     L^L4
        MLIN        23   3        W^W1     L^L3
        MLSC        24   W^W0     L^L3
        DEF2P       1    3        FS

OUT
    FS      DB[S21]    GR1
    FS      ANG[S21]   GR2
    FS      DB[S11]    GR3
    FS      DB[S22]    GR4

FREQ
    SWEEP       1        3       .05

GRID
    RANGE       1        3       .1
    GR1        -10       0        1
    GR2        -180     +180     20
    GR3        -50       0        5
```

Figure J.2 A typical Touchstone program to calculate the insertion loss, the return loss, and the phase of a cross-coupled transmission line.

Table J.1A

$\theta = 30°$

		$Z_1 = Z_0 = 35\Omega$		
FREQ-GHZ	DB[S21] PS	ANG[S21] PS	DB[S11] PS	DB[S22] PS
1.00000	-4.448	-20.197	-1.970	-1.970
1.05000	-3.792	-28.190	-2.390	-2.390
1.10000	-3.196	-36.414	-2.880	-2.880
1.15000	-2.664	-44.821	-3.441	-3.441
1.20000	-2.197	-53.351	-4.074	-4.074
1.25000	-1.796	-61.945	-4.774	-4.774
1.30000	-1.458	-70.542	-5.535	-5.535
1.35000	-1.178	-79.091	-6.344	-6.344
1.40000	-0.950	-87.550	-7.189	-7.189
1.45000	-0.767	-95.890	-8.053	-8.053
1.50000	-0.623	-104.093	-8.917	-8.917
1.55000	-0.510	-112.153	-9.766	-9.766
1.60000	-0.422	-120.073	-10.581	-10.581
1.65000	-0.355	-127.861	-11.349	-11.349
1.70000	-0.303	-135.531	-12.061	-12.061
1.75000	-0.263	-143.099	-12.713	-12.713
1.80000	-0.231	-150.583	-13.309	-13.309
1.85000	-0.206	-158.002	-13.860	-13.860
1.90000	-0.185	-165.378	-14.386	-14.386
1.95000	-0.167	-172.730	-14.916	-14.916
2.00000	-0.149	179.917	-15.490	-15.490
2.05000	-0.131	172.537	-16.161	-16.161
2.10000	-0.113	165.100	-17.005	-17.005
2.15000	-0.094	157.572	-18.141	-18.141
2.20000	-0.073	149.912	-19.775	-19.775
2.25000	-0.054	142.072	-22.348	-22.348
2.30000	-0.038	133.997	-27.189	-27.189
2.35000	-0.032	125.621	-48.031	-48.031
2.40000	-0.043	116.877	-26.779	-26.779
2.45000	-0.085	107.692	-19.567	-19.567
2.50000	-0.175	98.003	-15.121	-15.121
2.55000	-0.339	87.769	-11.817	-11.817
2.60000	-0.607	76.994	-9.184	-9.184
2.65000	-1.013	65.749	-7.037	-7.037
2.70000	-1.588	54.185	-5.287	-5.287
2.75000	-2.358	42.527	-3.884	-3.884
2.80000	-3.336	31.036	-2.787	-2.787
2.85000	-4.521	19.956	-1.954	-1.954
2.90000	-5.906	9.472	-1.338	-1.338
2.95000	-7.485	-0.315	-0.895	-0.895
3.00000	-9.263	-9.375	-0.583	-0.583

Table J.1A

$\theta = 30°$

$Z_1 = Z_0 = 32\Omega$

FREQ-GHZ	DB[S21] PS	ANG[S21] PS	DB[S11] PS	DB[S22] PS
1.00000	-4.137	-15.988	-2.156	-2.156
1.05000	-3.416	-24.400	-2.684	-2.684
1.10000	-2.771	-33.116	-3.317	-3.317
1.15000	-2.207	-42.057	-4.060	-4.060
1.20000	-1.728	-51.131	-4.915	-4.915
1.25000	-1.331	-60.244	-5.878	-5.878
1.30000	-1.013	-69.306	-6.937	-6.937
1.35000	-0.764	-78.245	-8.072	-8.072
1.40000	-0.575	-87.012	-9.258	-9.258
1.45000	-0.435	-95.576	-10.460	-10.460
1.50000	-0.332	-103.930	-11.642	-11.642
1.55000	-0.260	-112.080	-12.760	-12.760
1.60000	-0.209	-120.040	-13.769	-13.769
1.65000	-0.175	-127.833	-14.630	-14.630
1.70000	-0.152	-135.484	-15.311	-15.311
1.75000	-0.138	-143.015	-15.801	-15.801
1.80000	-0.130	-150.450	-16.105	-16.105
1.85000	-0.126	-157.810	-16.252	-16.252
1.90000	-0.126	-165.114	-16.282	-16.282
1.95000	-0.127	-172.382	-16.245	-16.245
2.00000	-0.128	-179.632	-16.195	-16.195
2.05000	-0.129	173.115	-16.185	-16.185
2.10000	-0.127	165.835	-16.270	-16.270
2.15000	-0.122	158.499	-16.514	-16.514
2.20000	-0.113	151.070	-16.999	-16.999
2.25000	-0.098	143.504	-17.850	-17.850
2.30000	-0.079	135.744	-19.289	-19.289
2.35000	-0.057	127.723	-21.821	-21.821
2.40000	-0.039	119.358	-27.101	-27.101
2.45000	-0.033	110.557	-46.522	-46.523
2.50000	-0.054	101.219	-23.693	-23.693
2.55000	-0.126	91.254	-17.032	-17.032
2.60000	-0.281	80.602	-12.746	-12.746
2.65000	-0.559	69.267	-9.554	-9.554
2.70000	-1.007	57.365	-7.064	-7.064
2.75000	-1.665	45.132	-5.114	-5.114
2.80000	-2.558	32.907	-3.614	-3.614
2.85000	-3.692	21.049	-2.494	-2.494
2.90000	-5.051	9.854	-1.683	-1.683
2.95000	-6.615	-0.501	-1.112	-1.112
3.00000	-8.370	-9.958	-0.719	-0.719

Table J.1B

$\theta = 30°$

$Z_1 = Z_0 = 30\Omega$				
FREQ-GHZ	DB[S21] PS	ANG[S21] PS	DB[S11] PS	DB[S22] PS
1.00000	-3.946	-13.988	-2.281	-2.281
1.05000	-3.187	-22.718	-2.889	-2.889
1.10000	-2.513	-31.799	-3.629	-3.629
1.15000	-1.935	-41.128	-4.513	-4.513
1.20000	-1.455	-50.585	-5.546	-5.546
1.25000	-1.070	-60.046	-6.723	-6.723
1.30000	-0.772	-69.404	-8.031	-8.031
1.35000	-0.551	-78.576	-9.445	-9.445
1.40000	-0.391	-87.510	-10.931	-10.931
1.45000	-0.279	-96.186	-12.440	-12.440
1.50000	-0.203	-104.604	-13.912	-13.912
1.55000	-0.153	-112.785	-15.270	-15.270
1.60000	-0.122	-120.756	-16.431	-16.431
1.65000	-0.103	-128.549	-17.314	-17.314
1.70000	-0.093	-136.194	-17.868	-17.868
1.75000	-0.089	-143.721	-18.082	-18.082
1.80000	-0.091	-151.154	-17.996	-17.996
1.85000	-0.096	-158.514	-17.683	-17.683
1.90000	-0.104	-165.818	-17.228	-17.229
1.95000	-0.115	-173.082	-16.715	-16.715
2.00000	-0.127	179.680	-16.211	-16.211
2.05000	-0.139	172.452	-15.771	-15.771
2.10000	-0.148	165.217	-15.443	-15.443
2.15000	-0.154	157.948	-15.268	-15.268
2.20000	-0.154	150.614	-15.296	-15.296
2.25000	-0.146	143.172	-15.591	-15.591
2.30000	-0.130	135.568	-16.257	-16.257
2.35000	-0.105	127.732	-17.488	-17.488
2.40000	-0.076	119.578	-19.707	-19.707
2.45000	-0.048	111.001	-24.211	-24.211
2.50000	-0.034	101.884	-43.959	-43.959
2.55000	-0.056	92.107	-23.644	-23.644
2.60000	-0.145	81.569	-16.244	-16.244
2.65000	-0.347	70.230	-11.745	-11.745
2.70000	-0.713	58.165	-8.505	-8.505
2.75000	-1.296	45.605	-6.064	-6.064
2.80000	-2.132	32.929	-4.229	-4.229
2.85000	-3.230	20.578	-2.882	-2.882
2.90000	-4.572	8.932	-1.923	-1.923
2.95000	-6.128	-1.775	-1.260	-1.260
3.00000	-7.871	-11.465	-0.811	-0.811

Table J.1B

$\theta = 30°$

$Z_1 = Z_0 = 27.5\Omega$			
FREQ-GHZ	DB[S21] PS	ANG[S21] PS	DB[S11] PS	DB[S22] PS
1.00000	-3.709	-10.494	-2.450	-2.450
1.05000	-2.890	-19.661	-3.186	-3.186
1.10000	-2.175	-29.258	-4.110	-4.110
1.15000	-1.577	-39.139	-5.247	-5.247
1.20000	-1.100	-49.131	-6.615	-6.615
1.25000	-0.738	-59.062	-8.220	-8.220
1.30000	-0.477	-68.792	-10.062	-10.062
1.35000	-0.299	-78.224	-12.129	-12.129
1.40000	-0.184	-87.313	-14.400	-14.400
1.45000	-0.113	-96.055	-16.826	-16.826
1.50000	-0.073	-104.476	-19.311	-19.311
1.55000	-0.051	-112.617	-21.668	-21.668
1.60000	-0.039	-120.529	-23.589	-23.589
1.65000	-0.035	-128.258	-24.703	-24.703
1.70000	-0.034	-135.846	-24.789	-24.789
1.75000	-0.037	-143.330	-23.963	-23.963
1.80000	-0.044	-150.737	-22.575	-22.575
1.85000	-0.055	-158.084	-20.970	-20.970
1.90000	-0.070	-165.385	-19.374	-19.374
1.95000	-0.091	-172.646	-17.903	-17.903
2.00000	-0.116	-179.872	-16.609	-16.609
2.05000	-0.144	172.932	-15.512	-15.512
2.10000	-0.174	165.759	-14.617	-14.617
2.15000	-0.201	158.597	-13.927	-13.927
2.20000	-0.223	151.424	-13.448	-13.448
2.25000	-0.236	144.207	-13.194	-13.194
2.30000	-0.237	136.899	-13.196	-13.196
2.35000	-0.223	129.440	-13.510	-13.510
2.40000	-0.193	121.745	-14.242	-14.242
2.45000	-0.149	113.712	-15.612	-15.612
2.50000	-0.097	105.213	-18.140	-18.140
2.55000	-0.052	96.098	-23.616	-23.616
2.60000	-0.037	86.214	-39.052	-39.051
2.65000	-0.091	75.426	-19.348	-19.348
2.70000	-0.270	63.684	-12.979	-12.979
2.75000	-0.639	51.094	-8.978	-8.978
2.80000	-1.262	37.984	-6.167	-6.167
2.85000	-2.174	24.865	-4.160	-4.160
2.90000	-3.370	12.291	-2.754	-2.754
2.95000	-4.814	0.677	-1.794	-1.794
3.00000	-6.457	-9.778	-1.154	-1.154

Table J.1C
$\theta = 30°$

$Z_1 = Z_0 = 25\Omega$				
FREQ-GHZ	DB[S21] PS	ANG[S21] PS	DB[S11] PS	DB[S22] PS
1.00000	-3.437	-6.383	-2.665	-2.665
1.05000	-2.544	-16.109	-3.589	-3.589
1.10000	-1.782	-26.387	-4.807	-4.807
1.15000	-1.169	-37.000	-6.382	-6.382
1.20000	-0.710	-47.689	-8.389	-8.389
1.25000	-0.393	-58.203	-10.924	-10.924
1.30000	-0.195	-68.354	-14.160	-14.160
1.35000	-0.085	-78.037	-18.481	-18.481
1.40000	-0.035	-87.226	-25.071	-25.071
1.45000	-0.021	-95.952	-44.104	-44.104
1.50000	-0.024	-104.280	-30.412	-30.412
1.55000	-0.033	-112.288	-25.284	-25.284
1.60000	-0.039	-120.055	-23.374	-23.374
1.65000	-0.042	-127.654	-22.906	-22.906
1.70000	-0.039	-135.145	-23.493	-23.493
1.75000	-0.032	-142.577	-25.198	-25.198
1.80000	-0.025	-149.984	-28.639	-28.639
1.85000	-0.020	-157.388	-36.969	-36.969
1.90000	-0.020	-164.801	-38.294	-38.294
1.95000	-0.027	-172.228	-27.813	-27.813
2.00000	-0.041	-179.665	-23.110	-23.110
2.05000	-0.062	172.888	-20.141	-20.141
2.10000	-0.088	165.431	-18.088	-18.088
2.15000	-0.116	157.954	-16.647	-16.647
2.20000	-0.140	150.435	-15.683	-15.683
2.25000	-0.157	142.832	-15.148	-15.148
2.30000	-0.161	135.083	-15.052	-15.052
2.35000	-0.150	127.097	-15.480	-15.480
2.40000	-0.121	118.752	-16.658	-16.658
2.45000	-0.082	109.883	-19.189	-19.189
2.50000	-0.045	100.291	-25.454	-25.454
2.55000	-0.039	89.747	-31.555	-31.555
2.60000	-0.117	78.038	-17.596	-17.596
2.65000	-0.359	65.066	-11.608	-11.608
2.70000	-0.863	50.991	-7.713	-7.713
2.75000	-1.719	36.345	-5.001	-5.001
2.80000	-2.968	21.935	-3.142	-3.142
2.85000	-4.583	8.532	-1.918	-1.918
2.90000	-6.496	-3.415	-1.145	-1.145
2.95000	-8.650	-13.811	-0.670	-0.670
3.00000	-11.023	-22.785	-0.384	-0.384

Table J.1C

$\theta = 30°$

$Z_1 = Z_0 = 22.5\Omega$				
FREQ-GHZ	DB[S21] PS	ANG[S21] PS	DB[S11] PS	DB[S22] PS
1.00000	-3.131	-3.131	-2.939	-2.939
1.05000	-2.166	-13.647	-4.125	-4.125
1.10000	-1.371	-24.845	-5.772	-5.772
1.15000	-0.774	-36.380	-8.037	-8.037
1.20000	-0.373	-47.858	-11.168	-11.168
1.25000	-0.143	-58.930	-15.721	-15.721
1.30000	-0.041	-69.377	-23.768	-23.768
1.35000	-0.022	-79.125	-35.519	-35.519
1.40000	-0.048	-88.206	-21.910	-21.910
1.45000	-0.090	-96.717	-17.939	-17.939
1.50000	-0.129	-104.778	-16.030	-16.030
1.55000	-0.154	-112.512	-15.115	-15.115
1.60000	-0.163	-120.028	-14.844	-14.844
1.65000	-0.155	-127.421	-15.090	-15.090
1.70000	-0.133	-134.765	-15.836	-15.836
1.75000	-0.102	-142.118	-17.156	-17.156
1.80000	-0.070	-149.515	-19.266	-19.266
1.85000	-0.041	-156.974	-22.734	-22.734
1.90000	-0.023	-164.497	-29.714	-29.714
1.95000	-0.018	-172.071	-39.555	-39.555
2.00000	-0.031	-179.672	-25.213	-25.213
2.05000	-0.061	172.724	-20.128	-20.128
2.10000	-0.104	165.141	-17.095	-17.095
2.15000	-0.156	157.590	-15.058	-15.057
2.20000	-0.211	150.069	-13.649	-13.649
2.25000	-0.259	142.551	-12.711	-12.711
2.30000	-0.293	134.984	-12.174	-12.174
2.35000	-0.303	127.285	-12.030	-12.030
2.40000	-0.285	119.330	-12.335	-12.335
2.45000	-0.237	110.952	-13.249	-13.249
2.50000	-0.162	101.930	-15.182	-15.182
2.55000	-0.082	91.986	-19.430	-19.430
2.60000	-0.037	80.804	-39.460	-39.460
2.65000	-0.105	68.115	-18.458	-18.458
2.70000	-0.401	53.872	-11.092	-11.092
2.75000	-1.061	38.495	-6.866	-6.866
2.80000	-2.171	22.933	-4.169	-4.169
2.85000	-3.722	8.301	-2.469	-2.469
2.90000	-5.624	-4.651	-1.437	-1.437
2.95000	-7.776	-15.718	-0.826	-0.826
3.00000	-10.117	-25.061	-0.470	-0.470

Table J.2A
$Z_1 = 32\Omega,\ \theta = 30°$

$Z_0 = 45\Omega$				
FREQ-GHZ	DB[S21] PS	ANG[S21] PS	DB[S11] PS	DB[S22] PS
1.00000	-1.958	-30.523	-4.474	-4.474
1.05000	-1.486	-39.377	-5.470	-5.470
1.10000	-1.101	-48.239	-6.613	-6.613
1.15000	-0.798	-57.017	-7.901	-7.901
1.20000	-0.567	-65.637	-9.326	-9.326
1.25000	-0.397	-74.046	-10.872	-10.872
1.30000	-0.275	-82.217	-12.519	-12.519
1.35000	-0.190	-90.142	-14.232	-14.232
1.40000	-0.134	-97.831	-15.964	-15.964
1.45000	-0.097	-105.302	-17.642	-17.642
1.50000	-0.075	-112.581	-19.168	-19.168
1.55000	-0.061	-119.695	-20.419	-20.419
1.60000	-0.054	-126.672	-21.278	-21.278
1.65000	-0.051	-133.537	-21.668	-21.668
1.70000	-0.051	-140.312	-21.588	-21.588
1.75000	-0.055	-147.013	-21.117	-21.117
1.80000	-0.061	-153.656	-20.373	-20.373
1.85000	-0.070	-160.250	-19.478	-19.478
1.90000	-0.083	-166.805	-18.530	-18.530
1.95000	-0.098	-173.325	-17.599	-17.599
2.00000	-0.115	-179.816	-16.732	-16.732
2.05000	-0.134	173.718	-15.955	-15.955
2.10000	-0.153	167.270	-15.289	-15.289
2.15000	-0.171	160.833	-14.745	-14.745
2.20000	-0.186	154.394	-14.334	-14.334
2.25000	-0.197	147.937	-14.070	-14.070
2.30000	-0.202	141.441	-13.969	-13.969
2.35000	-0.199	134.877	-14.060	-14.060
2.40000	-0.187	128.209	-14.383	-14.383
2.45000	-0.166	121.392	-15.012	-15.012
2.50000	-0.137	114.370	-16.075	-16.075
2.55000	-0.102	107.080	-17.832	-17.832
2.60000	-0.067	99.443	-20.925	-20.925
2.65000	-0.041	91.375	-27.881	-27.881
2.70000	-0.039	82.788	-33.156	-33.156
2.75000	-0.081	73.603	-20.219	-20.219
2.80000	-0.198	63.767	-14.574	-14.574
2.85000	-0.425	53.286	-10.809	-10.809
2.90000	-0.803	42.250	-8.009	-8.009
2.95000	-1.370	30.855	-5.857	-5.857
3.00000	-2.150	19.381	-4.205	-4.205

Table J.2A
$Z_1 = 32\Omega$, $\theta = 30°$

$Z_0 = 53\Omega$			
FREQ-GHZ	DB[S21]	ANG[S21]	DB[S11]	DB[S22]
PS	PS	PS	PS

FREQ-GHZ	DB[S21] PS	ANG[S21] PS	DB[S11] PS	DB[S22] PS
1.00000	-1.226	-37.909	-6.200	-6.200
1.05000	-0.882	-46.655	-7.502	-7.502
1.10000	-0.617	-55.266	-8.977	-8.978
1.15000	-0.420	-63.677	-10.627	-10.627
1.20000	-0.280	-71.847	-12.450	-12.450
1.25000	-0.183	-79.757	-14.440	-14.440
1.30000	-0.119	-87.409	-16.585	-16.585
1.35000	-0.079	-94.818	-18.855	-18.855
1.40000	-0.055	-102.007	-21.184	-21.184
1.45000	-0.041	-109.007	-23.441	-23.441
1.50000	-0.034	-115.846	-25.399	-25.399
1.55000	-0.030	-122.554	-26.748	-26.748
1.60000	-0.029	-129.154	-27.220	-27.220
1.65000	-0.030	-135.671	-26.775	-26.775
1.70000	-0.033	-142.121	-25.637	-25.637
1.75000	-0.038	-148.519	-24.122	-24.122
1.80000	-0.046	-154.873	-22.486	-22.486
1.85000	-0.057	-161.192	-20.883	-20.883
1.90000	-0.071	-167.477	-19.391	-19.391
1.95000	-0.090	-173.731	-18.042	-18.042
2.00000	-0.112	-179.954	-16.846	-16.846
2.05000	-0.137	173.854	-15.803	-15.803
2.10000	-0.164	167.694	-14.907	-14.907
2.15000	-0.192	161.561	-14.152	-14.152
2.20000	-0.219	155.452	-13.534	-13.534
2.25000	-0.244	149.358	-13.050	-13.050
2.30000	-0.263	143.266	-12.700	-12.700
2.35000	-0.276	137.161	-12.488	-12.488
2.40000	-0.280	131.019	-12.426	-12.426
2.45000	-0.275	124.812	-12.530	-12.530
2.50000	-0.259	118.504	-12.833	-12.833
2.55000	-0.231	112.052	-13.387	-13.387
2.60000	-0.194	105.404	-14.284	-14.284
2.65000	-0.149	98.501	-15.694	-15.694
2.70000	-0.102	91.272	-17.982	-17.982
2.75000	-0.060	83.643	-22.186	-22.186
2.80000	-0.038	75.534	-34.950	-34.950
2.85000	-0.053	66.872	-25.187	-25.187
2.90000	-0.130	57.608	-17.049	-17.049
2.95000	-0.301	47.735	-12.479	-12.479
3.00000	-0.600	37.312	-9.271	-9.271

Table J.2B

$Z_1 = 32\Omega, \ \theta = 30°$

		$Z_0 = 64\Omega$		
FREQ-GHZ	DB[S21] PS	ANG[S21] PS	DB[S11] PS	DB[S22] PS
1.00000	-0.695	-44.371	-8.482	-8.482
1.05000	-0.463	-52.804	-10.209	-10.209
1.10000	-0.297	-61.002	-12.190	-12.190
1.15000	-0.183	-68.934	-14.466	-14.466
1.20000	-0.109	-76.589	-17.104	-17.104
1.25000	-0.064	-83.975	-20.217	-20.217
1.30000	-0.039	-91.111	-24.022	-24.022
1.35000	-0.027	-98.026	-28.985	-28.985
1.40000	-0.022	-104.750	-36.510	-36.510
1.45000	-0.021	-111.316	-63.197	-63.197
1.50000	-0.021	-117.753	-42.169	-42.169
1.55000	-0.021	-124.088	-39.211	-39.211
1.60000	-0.021	-130.345	-40.605	-40.605
1.65000	-0.021	-136.544	-49.265	-49.265
1.70000	-0.021	-142.700	-44.075	-44.075
1.75000	-0.023	-148.824	-34.355	-34.355
1.80000	-0.026	-154.925	-29.175	-29.175
1.85000	-0.033	-161.006	-25.585	-25.585
1.90000	-0.044	-167.069	-22.856	-22.856
1.95000	-0.059	-173.114	-20.682	-20.682
2.00000	-0.078	-179.139	-18.908	-18.908
2.05000	-0.101	174.858	-17.440	-17.440
2.10000	-0.127	168.878	-16.221	-16.221
2.15000	-0.155	162.923	-15.210	-15.210
2.20000	-0.184	156.990	-14.381	-14.381
2.25000	-0.212	151.074	-13.716	-13.716
2.30000	-0.236	145.167	-13.203	-13.203
2.35000	-0.256	139.256	-12.838	-12.838
2.40000	-0.269	133.322	-12.622	-12.622
2.45000	-0.273	127.342	-12.563	-12.563
2.50000	-0.267	121.287	-12.680	-12.680
2.55000	-0.250	115.119	-13.006	-13.006
2.60000	-0.222	108.793	-13.597	-13.597
2.65000	-0.185	102.254	-14.554	-14.554
2.70000	-0.140	95.437	-16.072	-16.072
2.75000	-0.094	88.269	-18.583	-18.583
2.80000	-0.056	80.664	-23.425	-23.425
2.85000	-0.039	72.534	-45.053	-45.053
2.90000	-0.065	63.794	-22.934	-22.934
2.95000	-0.161	54.380	-15.859	-15.859
3.00000	-0.366	44.276	-11.583	-11.583

Table J.2B

$Z_1 = 32\Omega,\ \theta = 30°$

$Z_0 = 70\Omega$				
FREQ-GHZ	DB[S21] PS	ANG[S21] PS	DB[S11] PS	DB[S22] PS
1.00000	−0.497	−48.013	−9.914	−9.914
1.05000	−0.315	−56.264	−11.927	−11.927
1.10000	−0.191	−64.246	−14.274	−14.274
1.15000	−0.111	−71.943	−17.055	−17.055
1.20000	−0.063	−79.359	−20.455	−20.455
1.25000	−0.037	−86.511	−24.884	−24.884
1.30000	−0.025	−93.426	−31.554	−31.554
1.35000	−0.022	−100.135	−50.795	−50.794
1.40000	−0.022	−106.672	−36.988	−36.988
1.45000	−0.024	−113.068	−31.992	−31.992
1.50000	−0.025	−119.353	−30.282	−30.282
1.55000	−0.025	−125.551	−30.159	−30.159
1.60000	−0.024	−131.685	−31.392	−31.392
1.65000	−0.022	−137.773	−34.528	−34.528
1.70000	−0.021	−143.829	−43.044	−43.044
1.75000	−0.021	−149.862	−42.153	−42.153
1.80000	−0.024	−155.879	−31.974	−31.974
1.85000	−0.030	−161.881	−27.005	−27.005
1.90000	−0.040	−167.871	−23.662	−23.662
1.95000	−0.055	−173.845	−21.157	−21.157
2.00000	−0.074	−179.801	−19.182	−19.182
2.05000	−0.098	174.265	−17.583	−17.583
2.10000	−0.126	168.354	−16.270	−16.270
2.15000	−0.156	162.470	−15.190	−15.190
2.20000	−0.187	156.610	−14.305	−14.305
2.25000	−0.217	150.772	−13.590	−13.590
2.30000	−0.245	144.947	−13.032	−13.032
2.35000	−0.268	139.125	−12.620	−12.620
2.40000	−0.285	133.289	−12.353	−12.353
2.45000	−0.293	127.418	−12.234	−12.234
2.50000	−0.291	121.485	−12.277	−12.277
2.55000	−0.278	115.454	−12.505	−12.505
2.60000	−0.254	109.284	−12.963	−12.963
2.65000	−0.218	102.924	−13.722	−13.722
2.70000	−0.173	96.314	−14.918	−14.918
2.75000	−0.124	89.381	−16.826	−16.826
2.80000	−0.078	82.044	−20.136	−20.136
2.85000	−0.046	74.213	−27.727	−27.727
2.90000	−0.046	65.795	−30.529	−30.529
2.95000	−0.103	56.710	−18.846	−18.846
3.00000	−0.253	46.909	−13.451	−13.451

Table J.2C

$Z_1 = 32\Omega,\ \theta = 30°$

		$Z_0 = 76\Omega$		
FREQ-GHZ	DB[S21]	ANG[S21]	DB[S11]	DB[S22]
	PS	PS	PS	PS
1.00000	-0.372	-50.561	-11.192	-11.192
1.05000	-0.226	-58.637	-13.488	-13.488
1.10000	-0.130	-66.426	-16.232	-16.232
1.15000	-0.071	-73.924	-19.629	-19.629
1.20000	-0.040	-81.142	-24.150	-24.150
1.25000	-0.025	-88.105	-31.319	-31.319
1.30000	-0.022	-94.843	-66.196	-66.197
1.35000	-0.023	-101.390	-33.404	-33.404
1.40000	-0.027	-107.777	-28.882	-28.882
1.45000	-0.030	-114.036	-26.980	-26.980
1.50000	-0.031	-120.196	-26.327	-26.327
1.55000	-0.030	-126.281	-26.570	-26.570
1.60000	-0.028	-132.312	-27.695	-27.695
1.65000	-0.025	-138.305	-30.000	-30.000
1.70000	-0.023	-144.273	-34.626	-34.626
1.75000	-0.021	-150.225	-51.842	-51.842
1.80000	-0.022	-156.165	-35.902	-35.902
1.85000	-0.027	-162.096	-28.741	-28.741
1.90000	-0.036	-168.018	-24.648	-24.648
1.95000	-0.050	-173.926	-21.776	-21.776
2.00000	-0.070	-179.818	-19.588	-19.588
2.05000	-0.094	174.311	-17.850	-17.850
2.10000	-0.122	168.464	-16.440	-16.440
2.15000	-0.153	162.644	-15.285	-15.285
2.20000	-0.185	156.851	-14.339	-14.339
2.25000	-0.218	151.082	-13.573	-13.573
2.30000	-0.249	145.332	-12.965	-12.965
2.35000	-0.275	139.592	-12.505	-12.505
2.40000	-0.295	133.846	-12.186	-12.186
2.45000	-0.308	128.076	-12.008	-12.008
2.50000	-0.310	122.256	-11.981	-11.981
2.55000	-0.302	116.357	-12.120	-12.120
2.60000	-0.282	110.339	-12.459	-12.459
2.65000	-0.250	104.156	-13.050	-13.050
2.70000	-0.208	97.753	-13.988	-13.988
2.75000	-0.158	91.063	-15.453	-15.453
2.80000	-0.107	84.011	-17.837	-17.837
2.85000	-0.063	76.508	-22.292	-22.292
2.90000	-0.042	68.461	-37.604	-37.604
2.95000	-0.063	59.780	-23.746	-23.746
3.00000	-0.158	50.391	-16.093	-16.093

Table J.3A
$\theta = 45°$

$Z_1 = Z_0 = 35\Omega$			
FREQ-GHZ	DB[S21] PS	ANG[S21] PS	DB[S11] PS	DB[S22] PS
1.00000	-1.787	-34.562	-4.791	-4.791
1.05000	-1.365	-43.084	-5.781	-5.781
1.10000	-1.018	-51.594	-6.914	-6.914
1.15000	-0.741	-60.022	-8.195	-8.195
1.20000	-0.526	-68.311	-9.629	-9.629
1.25000	-0.365	-76.421	-11.223	-11.223
1.30000	-0.247	-84.328	-12.985	-12.985
1.35000	-0.164	-92.022	-14.929	-14.929
1.40000	-0.107	-99.508	-17.077	-17.077
1.45000	-0.070	-106.798	-19.459	-19.459
1.50000	-0.047	-113.909	-22.125	-22.125
1.55000	-0.034	-120.864	-25.147	-25.147
1.60000	-0.026	-127.684	-28.642	-28.642
1.65000	-0.022	-134.390	-32.815	-32.815
1.70000	-0.021	-141.004	-38.079	-38.079
1.75000	-0.020	-147.544	-45.556	-45.556
1.80000	-0.020	-154.026	-62.429	-62.427
1.85000	-0.020	-160.466	-55.687	-55.686
1.90000	-0.021	-166.875	-52.400	-52.400
1.95000	-0.021	-173.265	-53.712	-53.712
2.00000	-0.021	-179.646	-58.586	-58.586
2.05000	-0.021	173.972	-67.021	-67.020
2.10000	-0.022	167.578	-64.578	-64.576
2.15000	-0.022	161.164	-63.956	-63.956
2.20000	-0.022	154.717	-52.387	-52.387
2.25000	-0.023	148.225	-43.736	-43.736
2.30000	-0.024	141.674	-37.555	-37.555
2.35000	-0.026	135.048	-32.710	-32.710
2.40000	-0.031	128.327	-28.702	-28.702
2.45000	-0.039	121.491	-25.274	-25.274
2.50000	-0.052	114.519	-22.276	-22.276
2.55000	-0.075	107.387	-19.614	-19.614
2.60000	-0.112	100.073	-17.223	-17.223
2.65000	-0.168	92.559	-15.062	-15.062
2.70000	-0.250	84.830	-13.101	-13.101
2.75000	-0.367	76.882	-11.321	-11.321
2.80000	-0.527	68.721	-9.709	-9.709
2.85000	-0.741	60.371	-8.256	-8.256
2.90000	-1.019	51.872	-6.957	-6.957
2.95000	-1.367	43.282	-5.808	-5.808
3.00000	-1.793	34.671	-4.804	-4.804

Table J.3A
$\theta = 45°$

$Z_1 = Z_0 = 32\Omega$			
FREQ-GHZ	DB[S21] ANG[S21]		DB[S11]	DB[S22]

FREQ-GHZ	DB[S21] PS	ANG[S21] PS	DB[S11] PS	DB[S22] PS
1.00000	-1.479	-32.464	-5.482	-5.482
1.05000	-1.047	-41.553	-6.806	-6.806
1.10000	-0.710	-50.590	-8.376	-8.376
1.15000	-0.459	-59.474	-10.224	-10.224
1.20000	-0.281	-68.126	-12.400	-12.400
1.25000	-0.162	-76.499	-14.988	-14.988
1.30000	-0.088	-84.575	-18.149	-18.149
1.35000	-0.047	-92.358	-22.232	-22.232
1.40000	-0.027	-99.869	-28.222	-28.222
1.45000	-0.020	-107.137	-42.131	-42.131
1.50000	-0.021	-114.198	-35.847	-35.847
1.55000	-0.024	-121.086	-29.457	-29.457
1.60000	-0.028	-127.834	-26.856	-26.856
1.65000	-0.031	-134.474	-25.674	-25.674
1.70000	-0.032	-141.032	-25.325	-25.325
1.75000	-0.031	-147.530	-25.606	-25.606
1.80000	-0.029	-153.987	-26.468	-26.469
1.85000	-0.026	-160.415	-27.975	-27.975
1.90000	-0.024	-166.827	-30.348	-30.348
1.95000	-0.021	-173.231	-34.202	-34.202
2.00000	-0.020	-179.632	-42.044	-42.044
2.05000	-0.020	173.964	-48.502	-48.502
2.10000	-0.021	167.551	-36.584	-36.584
2.15000	-0.023	161.124	-32.164	-32.165
2.20000	-0.026	154.675	-29.736	-29.736
2.25000	-0.028	148.194	-28.397	-28.397
2.30000	-0.029	141.667	-27.892	-27.892
2.35000	-0.029	135.077	-28.227	-28.227
2.40000	-0.028	128.402	-29.679	-29.679
2.45000	-0.026	121.617	-33.251	-33.251
2.50000	-0.025	114.691	-45.916	-45.916
2.55000	-0.027	107.592	-34.673	-34.673
2.60000	-0.037	100.283	-26.242	-26.242
2.65000	-0.061	92.732	-21.248	-21.248
2.70000	-0.107	84.907	-17.561	-17.561
2.75000	-0.185	76.787	-14.602	-14.602
2.80000	-0.308	68.369	-12.130	-12.129
2.85000	-0.491	59.670	-10.026	-10.026
2.90000	-0.748	50.737	-8.225	-8.225
2.95000	-1.091	41.648	-6.687	-6.687
3.00000	-1.530	32.508	-5.386	-5.386

Table J.3B
$\theta = 45°$

$Z_1 = Z_0 = 30\Omega$

FREQ-GHZ	DB[S21] PS	ANG[S21] PS	DB[S11] PS	DB[S22] PS
1.00000	-1.303	-31.415	-5.960	-5.960
1.05000	-0.870	-40.876	-7.545	-7.545
1.10000	-0.546	-50.248	-9.480	-9.480
1.15000	-0.318	-59.404	-11.845	-11.845
1.20000	-0.169	-68.255	-14.792	-14.792
1.25000	-0.081	-76.755	-18.642	-18.642
1.30000	-0.037	-84.894	-24.298	-24.298
1.35000	-0.021	-92.689	-36.789	-36.789
1.40000	-0.022	-100.174	-32.815	-32.815
1.45000	-0.031	-107.395	-25.743	-25.743
1.50000	-0.042	-114.397	-22.791	-22.791
1.55000	-0.051	-121.224	-21.278	-21.278
1.60000	-0.057	-127.918	-20.558	-20.558
1.65000	-0.059	-134.514	-20.390	-20.390
1.70000	-0.056	-141.042	-20.676	-20.676
1.75000	-0.050	-147.525	-21.395	-21.395
1.80000	-0.043	-153.983	-22.588	-22.588
1.85000	-0.035	-160.427	-24.378	-24.378
1.90000	-0.028	-166.866	-27.060	-27.060
1.95000	-0.022	-173.306	-31.437	-31.437
2.00000	-0.020	-179.748	-41.468	-41.468
2.05000	-0.020	173.806	-40.021	-40.021
2.10000	-0.023	167.357	-31.141	-31.141
2.15000	-0.028	160.901	-27.113	-27.113
2.20000	-0.035	154.435	-24.680	-24.680
2.25000	-0.042	147.951	-23.124	-23.124
2.30000	-0.047	141.438	-22.188	-22.188
2.35000	-0.050	134.877	-21.779	-21.779
2.40000	-0.050	128.247	-21.899	-21.899
2.45000	-0.046	121.519	-22.656	-22.656
2.50000	-0.040	114.658	-24.346	-24.346
2.55000	-0.032	107.626	-27.840	-27.840
2.60000	-0.027	100.377	-37.908	-37.908
2.65000	-0.030	92.867	-32.427	-32.427
2.70000	-0.050	85.053	-23.169	-23.169
2.75000	-0.098	76.901	-18.105	-18.105
2.80000	-0.190	68.393	-14.492	-14.492
2.85000	-0.342	59.539	-11.668	-11.668
2.90000	-0.571	50.385	-9.373	-9.373
2.95000	-0.896	41.018	-7.482	-7.482
3.00000	-1.327	31.563	-5.925	-5.925

Table J.3B
$\theta = 45°$

		$Z_1 = Z_0 = 27.5\Omega$		
FREQ-GHZ	DB[S21] PS	ANG[S21] PS	DB[S11] PS	DB[S22] PS
1.00000	−1.058	−30.010	−6.770	−6.770
1.05000	−0.632	−40.057	−8.867	−8.867
1.10000	−0.338	−49.933	−11.586	−11.586
1.15000	−0.155	−59.473	−15.245	−15.245
1.20000	−0.059	−68.575	−20.734	−20.734
1.25000	−0.023	−77.200	−33.006	−33.006
1.30000	−0.025	−85.362	−29.074	−29.074
1.35000	−0.048	−93.106	−21.824	−21.824
1.40000	−0.078	−100.492	−18.705	−18.705
1.45000	−0.106	−107.588	−17.002	−17.002
1.50000	−0.128	−114.458	−16.053	−16.053
1.55000	−0.140	−121.162	−15.602	−15.602
1.60000	−0.142	−127.751	−15.532	−15.532
1.65000	−0.134	−134.268	−15.792	−15.792
1.70000	−0.120	−140.748	−16.372	−16.372
1.75000	−0.100	−147.215	−17.296	−17.296
1.80000	−0.078	−153.689	−18.633	−18.633
1.85000	−0.057	−160.179	−20.525	−20.525
1.90000	−0.039	−166.690	−23.276	−23.276
1.95000	−0.026	−173.219	−27.704	−27.704
2.00000	−0.019	−179.762	−37.837	−37.837
2.05000	−0.020	173.690	−36.122	−36.122
2.10000	−0.027	167.144	−27.273	−27.273
2.15000	−0.040	160.606	−23.189	−23.189
2.20000	−0.056	154.080	−20.655	−20.655
2.25000	−0.075	147.564	−18.946	−18.946
2.30000	−0.093	141.047	−17.790	−17.790
2.35000	−0.106	134.515	−17.062	−17.062
2.40000	−0.114	127.943	−16.713	−16.713
2.45000	−0.114	121.299	−16.744	−16.744
2.50000	−0.106	114.542	−17.207	−17.207
2.55000	−0.089	107.622	−18.239	−18.239
2.60000	−0.067	100.483	−20.157	−20.157
2.65000	−0.044	93.062	−23.854	−23.854
2.70000	−0.029	85.296	−34.207	−34.207
2.75000	−0.035	77.127	−28.654	−28.654
2.80000	−0.079	68.512	−19.593	−19.593
2.85000	−0.183	59.443	−14.712	−14.712
2.90000	−0.371	49.958	−11.304	−11.304
2.95000	−0.666	40.155	−8.720	−8.720
3.00000	−1.088	30.195	−6.701	−6.701

Table J.3C

$\theta = 45°$

$Z_1 = Z_0 = 25\Omega$			
FREQ-GHZ	DB[S21] PS	ANG[S21] PS	DB[S11] PS	DB[S22] PS
1.00000	-0.809	-29.012	-7.850	-7.850
1.05000	-0.408	-39.747	-10.764	-10.764
1.10000	-0.165	-50.175	-14.983	-14.983
1.15000	-0.048	-60.087	-22.281	-22.281
1.20000	-0.022	-69.381	-39.509	-39.509
1.25000	-0.050	-78.047	-21.638	-21.638
1.30000	-0.105	-86.136	-17.066	-17.066
1.35000	-0.168	-93.734	-14.711	-14.711
1.40000	-0.224	-100.936	-13.341	-13.341
1.45000	-0.266	-107.836	-12.549	-12.549
1.50000	-0.291	-114.518	-12.153	-12.153
1.55000	-0.296	-121.058	-12.067	-12.067
1.60000	-0.285	-127.516	-12.249	-12.249
1.65000	-0.258	-133.944	-12.689	-12.689
1.70000	-0.221	-140.381	-13.403	-13.403
1.75000	-0.177	-146.854	-14.431	-14.431
1.80000	-0.132	-153.380	-15.857	-15.857
1.85000	-0.090	-159.966	-17.839	-17.839
1.90000	-0.055	-166.608	-20.712	-20.712
1.95000	-0.030	-173.296	-25.397	-25.397
2.00000	-0.018	179.989	-37.015	-37.015
2.05000	-0.020	173.268	-31.829	-31.829
2.10000	-0.036	166.560	-23.835	-23.835
2.15000	-0.062	159.884	-19.925	-19.925
2.20000	-0.097	153.254	-17.439	-17.439
2.25000	-0.136	146.675	-15.721	-15.721
2.30000	-0.175	140.141	-14.508	-14.508
2.35000	-0.209	133.639	-13.676	-13.676
2.40000	-0.234	127.142	-13.160	-13.160
2.45000	-0.247	120.614	-12.937	-12.937
2.50000	-0.244	114.005	-13.014	-13.014
2.55000	-0.224	107.257	-13.435	-13.435
2.60000	-0.188	100.299	-14.299	-14.299
2.65000	-0.140	93.048	-15.818	-15.818
2.70000	-0.088	85.419	-18.502	-18.502
2.75000	-0.045	77.323	-24.088	-24.088
2.80000	-0.030	68.684	-41.217	-41.217
2.85000	-0.070	59.460	-20.589	-20.589
2.90000	-0.197	49.669	-14.359	-14.359
2.95000	-0.444	39.414	-10.493	-10.493
3.00000	-0.840	28.892	-7.750	-7.750

Table J.3C

$\theta = 45°$

$Z_1 = Z_0 = 22.5\Omega$			
FREQ-GHZ	DB[S21] PS	ANG[S21] PS	DB[S11] PS	DB[S22] PS
1.00000	-0.544	-28.476	-9.519	-9.519
1.05000	-0.199	-40.050	-14.095	-14.095
1.10000	-0.045	-51.076	-23.172	-23.172
1.15000	-0.029	-61.308	-27.497	-27.497
1.20000	-0.099	-70.674	-17.447	-17.447
1.25000	-0.209	-79.226	-13.693	-13.693
1.30000	-0.326	-87.083	-11.647	-11.647
1.35000	-0.431	-94.385	-10.423	-10.423
1.40000	-0.511	-101.267	-9.692	-9.692
1.45000	-0.560	-107.852	-9.299	-9.299
1.50000	-0.578	-114.247	-9.167	-9.167
1.55000	-0.566	-120.540	-9.258	-9.258
1.60000	-0.527	-126.804	-9.558	-9.558
1.65000	-0.466	-133.100	-10.076	-10.076
1.70000	-0.391	-139.472	-10.836	-10.836
1.75000	-0.308	-145.952	-11.889	-11.889
1.80000	-0.224	-152.556	-13.326	-13.326
1.85000	-0.147	-159.286	-15.310	-15.310
1.90000	-0.083	-166.130	-18.179	-18.179
1.95000	-0.039	-173.062	-22.857	-22.857
2.00000	-0.018	179.954	-34.452	-34.452
2.05000	-0.021	172.959	-29.305	-29.305
2.10000	-0.049	165.995	-21.302	-21.302
2.15000	-0.096	159.100	-17.388	-17.388
2.20000	-0.159	152.303	-14.895	-14.895
2.25000	-0.232	145.618	-13.161	-13.161
2.30000	-0.306	139.048	-11.919	-11.919
2.35000	-0.374	132.578	-11.035	-11.035
2.40000	-0.430	126.180	-10.437	-10.437
2.45000	-0.466	119.811	-10.089	-10.089
2.50000	-0.480	113.415	-9.979	-9.979
2.55000	-0.465	106.919	-10.118	-10.118
2.60000	-0.422	100.239	-10.549	-10.549
2.65000	-0.353	93.275	-11.359	-11.359
2.70000	-0.263	85.912	-12.728	-12.728
2.75000	-0.164	78.028	-15.058	-15.058
2.80000	-0.077	69.500	-19.537	-19.537
2.85000	-0.032	60.235	-35.691	-35.691
2.90000	-0.069	50.202	-20.831	-20.831
2.95000	-0.236	39.482	-13.463	-13.463
3.00000	-0.577	28.307	-9.340	-9.340

Table J.4A

$Z_1 = 32\Omega$, $\theta = 45°$

	$Z_0 = 45\Omega$			
FREQ-GHZ	DB[S21] PS	ANG[S21] PS	DB[S11] PS	DB[S22] PS
1.00000	−0.404	−47.983	−10.792	−10.792
1.05000	−0.231	−56.426	−13.312	−13.312
1.10000	−0.121	−64.569	−16.494	−16.494
1.15000	−0.057	−72.390	−20.815	−20.815
1.20000	−0.028	−79.893	−27.876	−27.876
1.25000	−0.020	−87.098	−53.019	−53.019
1.30000	−0.026	−94.037	−28.615	−28.615
1.35000	−0.037	−100.745	−23.785	−23.785
1.40000	−0.051	−107.259	−21.355	−21.355
1.45000	−0.063	−113.615	−19.975	−19.975
1.50000	−0.071	−119.848	−19.216	−19.216
1.55000	−0.076	−125.986	−18.895	−18.895
1.60000	−0.075	−132.055	−18.926	−18.926
1.65000	−0.071	−138.077	−19.274	−19.274
1.70000	−0.064	−144.070	−19.938	−19.938
1.75000	−0.055	−150.046	−20.952	−20.952
1.80000	−0.045	−156.015	−22.395	−22.395
1.85000	−0.035	−161.983	−24.428	−24.428
1.90000	−0.028	−167.955	−27.413	−27.413
1.95000	−0.022	−173.930	−32.389	−32.389
2.00000	−0.020	−179.910	−46.259	−46.259
2.05000	−0.021	174.109	−36.872	−36.872
2.10000	−0.025	168.127	−29.621	−29.621
2.15000	−0.032	162.145	−25.899	−25.899
2.20000	−0.040	156.162	−23.503	−23.503
2.25000	−0.050	150.176	−21.845	−21.845
2.30000	−0.059	144.180	−20.692	−20.692
2.35000	−0.066	138.166	−19.932	−19.932
2.40000	−0.071	132.121	−19.519	−19.519
2.45000	−0.073	126.028	−19.446	−19.446
2.50000	−0.070	119.866	−19.745	−19.745
2.55000	−0.063	113.611	−20.504	−20.504
2.60000	−0.053	107.233	−21.918	−21.918
2.65000	−0.042	100.700	−24.452	−24.452
2.70000	−0.032	93.975	−29.646	−29.646
2.75000	−0.029	87.024	−52.917	−52.918
2.80000	−0.038	79.810	−27.271	−27.271
2.85000	−0.070	72.303	−20.612	−20.612
2.90000	−0.134	64.484	−16.416	−16.416
2.95000	−0.245	56.349	−13.292	−13.292
3.00000	−0.416	47.918	−10.803	−10.803

Table J.4A

$Z_1 = 32\Omega$, $\theta = 45°$

	$Z_0 = 53\Omega$			
FREQ-GHZ	DB[S21] PS	ANG[S21] PS	DB[S11] PS	DB[S22] PS
1.00000	-0.164	-54.576	-14.967	-14.967
1.05000	-0.078	-62.498	-18.916	-18.916
1.10000	-0.035	-70.086	-25.050	-25.050
1.15000	-0.020	-77.351	-43.552	-43.552
1.20000	-0.025	-84.320	-29.204	-29.204
1.25000	-0.040	-91.026	-23.305	-23.305
1.30000	-0.059	-97.505	-20.449	-20.449
1.35000	-0.077	-103.797	-18.783	-18.783
1.40000	-0.092	-109.937	-17.778	-17.778
1.45000	-0.102	-115.957	-17.215	-17.215
1.50000	-0.107	-121.888	-16.984	-16.984
1.55000	-0.106	-127.754	-17.033	-17.033
1.60000	-0.100	-133.578	-17.338	-17.338
1.65000	-0.090	-139.377	-17.899	-17.899
1.70000	-0.078	-145.165	-18.739	-18.739
1.75000	-0.064	-150.951	-19.910	-19.910
1.80000	-0.051	-156.742	-21.510	-21.510
1.85000	-0.038	-162.542	-23.730	-23.730
1.90000	-0.029	-168.351	-27.010	-27.010
1.95000	-0.022	-174.167	-32.722	-32.722
2.00000	-0.020	-179.989	-59.112	-59.112
2.05000	-0.022	174.189	-33.555	-33.555
2.10000	-0.028	168.369	-27.410	-27.410
2.15000	-0.038	162.554	-23.979	-23.979
2.20000	-0.050	156.747	-21.677	-21.677
2.25000	-0.065	150.948	-20.024	-20.024
2.30000	-0.079	145.152	-18.812	-18.812
2.35000	-0.092	139.355	-17.938	-17.938
2.40000	-0.103	133.547	-17.346	-17.346
2.45000	-0.110	127.714	-17.011	-17.011
2.50000	-0.112	121.841	-16.930	-16.930
2.55000	-0.109	115.906	-17.123	-17.123
2.60000	-0.100	109.884	-17.640	-17.640
2.65000	-0.086	103.746	-18.577	-18.577
2.70000	-0.069	97.462	-20.133	-20.133
2.75000	-0.051	90.994	-22.769	-22.769
2.80000	-0.036	84.307	-27.971	-27.971
2.85000	-0.030	77.363	-53.742	-53.742
2.90000	-0.042	70.131	-26.266	-26.266
2.95000	-0.082	62.584	-19.594	-19.594
3.00000	-0.162	54.710	-15.455	-15.455

Table J.4B

$Z_1 = 32\Omega$, $\theta = 45°$

$Z_0 = 64\Omega$				
FREQ-GHZ	DB[S21] PS	ANG[S21] PS	DB[S11] PS	DB[S22] PS
1.00000	-0.048	-60.164	-22.248	-22.248
1.05000	-0.023	-67.518	-33.627	-33.627
1.10000	-0.023	-74.547	-31.223	-31.223
1.15000	-0.040	-81.280	-23.360	-23.360
1.20000	-0.064	-87.754	-19.915	-19.915
1.25000	-0.090	-94.007	-17.910	-17.910
1.30000	-0.114	-100.076	-16.647	-16.647
1.35000	-0.133	-105.996	-15.850	-15.850
1.40000	-0.147	-111.800	-15.384	-15.384
1.45000	-0.153	-117.518	-15.180	-15.180
1.50000	-0.153	-123.174	-15.197	-15.197
1.55000	-0.146	-128.791	-15.418	-15.418
1.60000	-0.134	-134.387	-15.841	-15.841
1.65000	-0.119	-139.975	-16.475	-16.475
1.70000	-0.100	-145.568	-17.349	-17.349
1.75000	-0.081	-151.172	-18.511	-18.511
1.80000	-0.063	-156.792	-20.047	-20.047
1.85000	-0.047	-162.430	-22.119	-22.119
1.90000	-0.034	-168.084	-25.067	-25.067
1.95000	-0.025	-173.751	-29.828	-29.828
2.00000	-0.021	-179.428	-41.749	-41.749
2.05000	-0.022	174.891	-35.924	-35.924
2.10000	-0.027	169.212	-28.017	-28.017
2.15000	-0.038	163.539	-24.069	-24.069
2.20000	-0.052	157.876	-21.501	-21.501
2.25000	-0.068	152.223	-19.668	-19.668
2.30000	-0.086	146.582	-18.310	-18.310
2.35000	-0.103	140.946	-17.299	-17.299
2.40000	-0.119	135.311	-16.564	-16.564
2.45000	-0.131	129.665	-16.069	-16.069
2.50000	-0.139	123.995	-15.796	-15.796
2.55000	-0.141	118.282	-15.746	-15.746
2.60000	-0.137	112.505	-15.935	-15.935
2.65000	-0.126	106.637	-16.406	-16.406
2.70000	-0.109	100.650	-17.238	-17.238
2.75000	-0.088	94.509	-18.583	-18.583
2.80000	-0.065	88.178	-20.767	-20.767
2.85000	-0.045	81.617	-24.684	-24.684
2.90000	-0.033	74.787	-35.118	-35.118
2.95000	-0.038	67.649	-29.999	-29.999
3.00000	-0.070	60.173	-21.024	-21.024

Table J.4B

$Z_1 = 32\Omega$, $\theta = 45°$

$Z_0 = 70\Omega$				
FREQ-GHZ	DB[S21] PS	ANG[S21] PS	DB[S11] PS	DB[S22] PS
1.00000	-0.027	-62.762	-29.292	-29.292
1.05000	-0.022	-69.863	-35.124	-35.124
1.10000	-0.036	-76.652	-24.237	-24.237
1.15000	-0.061	-83.165	-20.201	-20.201
1.20000	-0.090	-89.439	-17.926	-17.926
1.25000	-0.118	-95.512	-16.493	-16.493
1.30000	-0.141	-101.420	-15.567	-15.567
1.35000	-0.159	-107.199	-14.993	-14.993
1.40000	-0.169	-112.878	-14.685	-14.685
1.45000	-0.173	-118.485	-14.599	-14.599
1.50000	-0.169	-124.043	-14.709	-14.709
1.55000	-0.159	-129.572	-15.007	-15.007
1.60000	-0.144	-135.089	-15.495	-15.495
1.65000	-0.125	-140.607	-16.190	-16.190
1.70000	-0.105	-146.135	-17.124	-17.124
1.75000	-0.084	-151.680	-18.351	-18.351
1.80000	-0.064	-157.244	-19.967	-19.967
1.85000	-0.047	-162.828	-22.152	-22.152
1.90000	-0.033	-168.431	-25.300	-25.300
1.95000	-0.024	-174.047	-30.562	-30.562
2.00000	-0.021	-179.673	-47.007	-47.007
2.05000	-0.022	174.698	-33.639	-33.639
2.10000	-0.030	169.072	-26.817	-26.817
2.15000	-0.042	163.456	-23.152	-23.152
2.20000	-0.058	157.852	-20.706	-20.706
2.25000	-0.077	152.263	-18.933	-18.933
2.30000	-0.098	146.690	-17.600	-17.600
2.35000	-0.118	141.130	-16.591	-16.591
2.40000	-0.137	135.576	-15.841	-15.841
2.45000	-0.152	130.020	-15.311	-15.311
2.50000	-0.163	124.449	-14.983	-14.983
2.55000	-0.168	118.846	-14.851	-14.851
2.60000	-0.166	113.192	-14.925	-14.925
2.65000	-0.157	107.461	-15.230	-15.230
2.70000	-0.141	101.627	-15.814	-15.814
2.75000	-0.119	95.657	-16.768	-16.768
2.80000	-0.094	89.516	-18.263	-18.263
2.85000	-0.067	83.166	-20.676	-20.676
2.90000	-0.045	76.566	-25.095	-25.095
2.95000	-0.033	69.677	-39.103	-39.103
3.00000	-0.043	62.461	-27.495	-27.495

Table J.4C

$Z_1 = 32\Omega, \ \theta = 45°$

$Z_0 = 76\Omega$				
FREQ-GHZ	DB[S21] PS	ANG[S21] PS	DB[S11] PS	DB[S22] PS
1.00000	-0.021	-64.736	-51.567	-51.566
1.05000	-0.030	-71.644	-26.811	-26.811
1.10000	-0.052	-78.255	-21.300	-21.300
1.15000	-0.082	-84.604	-18.471	-18.471
1.20000	-0.112	-90.732	-16.734	-16.734
1.25000	-0.140	-96.675	-15.607	-15.607
1.30000	-0.163	-102.468	-14.879	-14.879
1.35000	-0.178	-108.146	-14.442	-14.442
1.40000	-0.186	-113.736	-14.236	-14.236
1.45000	-0.187	-119.264	-14.230	-14.230
1.50000	-0.180	-124.754	-14.405	-14.405
1.55000	-0.167	-130.222	-14.760	-14.760
1.60000	-0.150	-135.685	-15.300	-15.300
1.65000	-0.129	-141.153	-16.046	-16.046
1.70000	-0.107	-146.637	-17.034	-17.034
1.75000	-0.084	-152.140	-18.326	-18.326
1.80000	-0.064	-157.666	-20.026	-20.026
1.85000	-0.046	-163.213	-22.341	-22.341
1.90000	-0.032	-168.780	-25.731	-25.731
1.95000	-0.023	-174.361	-31.672	-31.672
2.00000	-0.021	-179.950	-67.377	-67.376
2.05000	-0.024	174.460	-31.657	-31.657
2.10000	-0.032	168.875	-25.711	-25.711
2.15000	-0.046	163.301	-22.313	-22.313
2.20000	-0.065	157.743	-19.990	-19.990
2.25000	-0.086	152.205	-18.281	-18.281
2.30000	-0.109	146.687	-16.982	-16.982
2.35000	-0.133	141.187	-15.985	-15.985
2.40000	-0.154	135.700	-15.230	-15.230
2.45000	-0.173	130.218	-14.681	-14.681
2.50000	-0.187	124.729	-14.318	-14.318
2.55000	-0.195	119.218	-14.133	-14.133
2.60000	-0.196	113.667	-14.129	-14.129
2.65000	-0.189	108.052	-14.323	-14.323
2.70000	-0.174	102.349	-14.746	-14.746
2.75000	-0.152	96.526	-15.458	-15.458
2.80000	-0.125	90.552	-16.562	-16.562
2.85000	-0.094	84.389	-18.265	-18.265
2.90000	-0.065	77.999	-21.031	-21.031
2.95000	-0.042	71.342	-26.362	-26.362
3.00000	-0.034	64.379	-52.034	-52.034

Table J.5A
$\theta = 60°$

$Z_1 = Z_0 = 35\Omega$			
FREQ-GHZ	DB[S21] PS	ANG[S21] PS	DB[S11] PS	DB[S22] PS
1.00000	-0.596	-46.869	-9.102	-9.102
1.05000	-0.379	-55.225	-11.056	-11.056
1.10000	-0.225	-63.365	-13.416	-13.416
1.15000	-0.123	-71.254	-16.360	-16.360
1.20000	-0.061	-78.879	-20.285	-20.285
1.25000	-0.030	-86.242	-26.392	-26.392
1.30000	-0.019	-93.357	-45.200	-45.201
1.35000	-0.023	-100.246	-30.258	-30.258
1.40000	-0.035	-106.935	-24.290	-24.290
1.45000	-0.051	-113.451	-21.317	-21.317
1.50000	-0.068	-119.821	-19.501	-19.501
1.55000	-0.084	-126.073	-18.312	-18.312
1.60000	-0.097	-132.230	-17.520	-17.520
1.65000	-0.107	-138.313	-17.008	-17.008
1.70000	-0.113	-144.342	-16.703	-16.703
1.75000	-0.117	-150.333	-16.560	-16.560
1.80000	-0.117	-156.300	-16.546	-16.546
1.85000	-0.115	-162.255	-16.636	-16.636
1.90000	-0.112	-168.208	-16.808	-16.808
1.95000	-0.107	-174.167	-17.041	-17.041
2.00000	-0.102	179.860	-17.315	-17.315
2.05000	-0.097	173.868	-17.605	-17.605
2.10000	-0.092	167.852	-17.883	-17.883
2.15000	-0.089	161.804	-18.119	-18.119
2.20000	-0.087	155.720	-18.278	-18.278
2.25000	-0.086	149.593	-18.325	-18.325
2.30000	-0.088	143.416	-18.230	-18.230
2.35000	-0.093	137.181	-17.969	-17.969
2.40000	-0.101	130.877	-17.530	-17.530
2.45000	-0.113	124.496	-16.916	-16.916
2.50000	-0.132	118.024	-16.144	-16.144
2.55000	-0.157	111.451	-15.239	-15.239
2.60000	-0.193	104.765	-14.234	-14.234
2.65000	-0.242	97.955	-13.163	-13.163
2.70000	-0.308	91.013	-12.056	-12.056
2.75000	-0.394	83.936	-10.941	-10.941
2.80000	-0.506	76.724	-9.842	-9.842
2.85000	-0.650	69.385	-8.777	-8.777
2.90000	-0.830	61.935	-7.762	-7.762
2.95000	-1.052	54.400	-6.810	-6.810
3.00000	-1.319	46.812	-5.928	-5.928

Table J.5A

$\theta = 60°$

FREQ-GHZ	$Z_1 = Z_0 = 32\Omega$			
	DB[S21] PS	ANG[S21] PS	DB[S11] PS	DB[S22] PS
1.00000	-0.395	-46.491	-10.876	-10.876
1.05000	-0.209	-55.267	-13.763	-13.763
1.10000	-0.096	-63.721	-17.677	-17.677
1.15000	-0.038	-71.819	-23.848	-23.848
1.20000	-0.020	-79.557	-43.888	-43.888
1.25000	-0.027	-86.954	-27.379	-27.379
1.30000	-0.049	-94.043	-21.562	-21.562
1.35000	-0.078	-100.863	-18.671	-18.671
1.40000	-0.107	-107.457	-16.930	-16.930
1.45000	-0.134	-113.867	-15.820	-15.820
1.50000	-0.154	-120.130	-15.118	-15.118
1.55000	-0.168	-126.281	-14.710	-14.710
1.60000	-0.174	-132.352	-14.531	-14.531
1.65000	-0.174	-138.366	-14.543	-14.543
1.70000	-0.168	-144.347	-14.723	-14.723
1.75000	-0.156	-150.311	-15.059	-15.059
1.80000	-0.142	-156.271	-15.545	-15.545
1.85000	-0.125	-162.238	-16.181	-16.181
1.90000	-0.107	-168.219	-16.971	-16.971
1.95000	-0.090	-174.217	-17.922	-17.922
2.00000	-0.074	179.764	-19.044	-19.044
2.05000	-0.060	173.725	-20.348	-20.348
2.10000	-0.048	167.664	-21.845	-21.845
2.15000	-0.039	161.582	-23.538	-23.538
2.20000	-0.032	155.477	-25.409	-25.409
2.25000	-0.028	149.346	-27.390	-27.390
2.30000	-0.026	143.185	-29.313	-29.313
2.35000	-0.024	136.986	-30.838	-30.838
2.40000	-0.024	130.741	-31.483	-31.483
2.45000	-0.025	124.436	-30.896	-30.896
2.50000	-0.028	118.056	-29.179	-29.179
2.55000	-0.032	111.583	-26.784	-26.784
2.60000	-0.040	104.999	-24.148	-24.148
2.65000	-0.055	98.284	-21.534	-21.534
2.70000	-0.080	91.419	-19.062	-19.062
2.75000	-0.119	84.387	-16.774	-16.774
2.80000	-0.178	77.179	-14.681	-14.681
2.85000	-0.264	69.791	-12.778	-12.778
2.90000	-0.385	62.233	-11.056	-11.056
2.95000	-0.548	54.525	-9.507	-9.507
3.00000	-0.760	46.706	-8.124	-8.124

Table J.5B
$\theta = 60°$

$Z_1 = Z_0 = 30\Omega$

FREQ-GHZ	DB[S21] PS	ANG[S21] PS	DB[S11] PS	DB[S22] PS
1.00000	-0.290	-46.300	-12.262	-12.262
1.05000	-0.130	-55.327	-16.118	-16.118
1.10000	-0.047	-63.954	-22.241	-22.241
1.15000	-0.020	-72.149	-41.968	-41.968
1.20000	-0.030	-79.922	-25.834	-25.834
1.25000	-0.062	-87.303	-19.995	-19.995
1.30000	-0.104	-94.341	-17.102	-17.102
1.35000	-0.146	-101.090	-15.364	-15.364
1.40000	-0.184	-107.601	-14.259	-14.259
1.45000	-0.214	-113.926	-13.561	-13.561
1.50000	-0.233	-120.110	-13.158	-13.158
1.55000	-0.242	-126.192	-12.985	-12.985
1.60000	-0.241	-132.207	-13.007	-13.007
1.65000	-0.231	-138.184	-13.203	-13.203
1.70000	-0.214	-144.144	-13.565	-13.565
1.75000	-0.191	-150.106	-14.093	-14.093
1.80000	-0.165	-156.081	-14.797	-14.797
1.85000	-0.137	-162.080	-15.693	-15.693
1.90000	-0.110	-168.104	-16.809	-16.809
1.95000	-0.085	-174.156	-18.188	-18.188
2.00000	-0.063	179.766	-19.898	-19.898
2.05000	-0.046	173.666	-22.056	-22.056
2.10000	-0.033	167.548	-24.873	-24.873
2.15000	-0.025	161.415	-28.823	-28.823
2.20000	-0.020	155.272	-35.401	-35.401
2.25000	-0.019	149.117	-68.691	-68.692
2.30000	-0.020	142.949	-36.903	-36.903
2.35000	-0.023	136.761	-31.978	-31.978
2.40000	-0.025	130.545	-29.836	-29.836
2.45000	-0.026	124.286	-29.141	-29.141
2.50000	-0.026	117.967	-29.695	-29.695
2.55000	-0.025	111.566	-31.944	-31.944
2.60000	-0.023	105.060	-38.370	-38.370
2.65000	-0.024	98.422	-42.382	-42.382
2.70000	-0.029	91.626	-29.486	-29.486
2.75000	-0.045	84.647	-23.550	-23.550
2.80000	-0.076	77.465	-19.464	-19.464
2.85000	-0.131	70.070	-16.297	-16.297
2.90000	-0.218	62.462	-13.704	-13.704
2.95000	-0.348	54.658	-11.523	-11.523
3.00000	-0.529	46.695	-9.665	-9.665

Table J.5B

$\theta = 60°$

$Z_1 = Z_0 = 27.5\Omega$				
FREQ-GHZ	DB[S21] PS	ANG[S21] PS	DB[S11] PS	DB[S22] PS
1.00000	-0.158	-46.203	-15.156	-15.156
1.05000	-0.048	-55.588	-22.236	-22.236
1.10000	-0.021	-64.431	-41.353	-41.353
1.15000	-0.048	-72.715	-21.776	-21.776
1.20000	-0.107	-80.473	-16.972	-16.972
1.25000	-0.178	-87.767	-14.427	-14.427
1.30000	-0.249	-94.670	-12.864	-12.864
1.35000	-0.310	-101.256	-11.861	-11.861
1.40000	-0.358	-107.597	-11.225	-11.225
1.45000	-0.390	-113.755	-10.857	-10.857
1.50000	-0.404	-119.787	-10.701	-10.701
1.55000	-0.402	-125.740	-10.725	-10.725
1.60000	-0.385	-131.654	-10.913	-10.913
1.65000	-0.356	-137.560	-11.260	-11.260
1.70000	-0.317	-143.485	-11.771	-11.771
1.75000	-0.272	-149.445	-12.459	-12.459
1.80000	-0.224	-155.454	-13.347	-13.347
1.85000	-0.176	-161.517	-14.474	-14.474
1.90000	-0.131	-167.633	-15.900	-15.900
1.95000	-0.092	-173.798	-17.727	-17.727
2.00000	-0.060	179.997	-20.139	-20.139
2.05000	-0.037	173.765	-23.533	-23.533
2.10000	-0.023	167.516	-29.073	-29.073
2.15000	-0.018	161.264	-46.019	-46.019
2.20000	-0.020	155.018	-32.686	-32.686
2.25000	-0.028	148.784	-26.243	-26.243
2.30000	-0.040	142.565	-22.987	-22.987
2.35000	-0.053	136.356	-21.002	-21.002
2.40000	-0.065	130.148	-19.763	-19.763
2.45000	-0.074	123.924	-19.065	-19.065
2.50000	-0.077	117.664	-18.832	-18.832
2.55000	-0.075	111.339	-19.067	-19.067
2.60000	-0.066	104.918	-19.854	-19.854
2.65000	-0.054	98.363	-21.421	-21.421
2.70000	-0.039	91.637	-24.368	-24.368
2.75000	-0.028	84.700	-31.079	-31.079
2.80000	-0.026	77.518	-37.483	-37.483
2.85000	-0.044	70.064	-24.010	-24.010
2.90000	-0.092	62.325	-18.329	-18.329
2.95000	-0.184	54.311	-14.559	-14.559
3.00000	-0.333	46.059	-11.726	-11.726

Table J.5C

$\theta = 60°$

		$Z_1 = Z_0 = 25\Omega$		
FREQ-GHZ	DB[S21] PS	ANG[S21] PS	DB[S11] PS	DB[S22] PS
1.00000	-0.065	-46.503	-20.198	-20.198
1.05000	-0.022	-56.243	-40.431	-40.431
1.10000	-0.064	-65.259	-19.921	-19.921
1.15000	-0.156	-73.565	-15.087	-15.087
1.20000	-0.267	-81.235	-12.547	-12.547
1.25000	-0.378	-88.369	-10.996	-10.996
1.30000	-0.475	-95.071	-10.000	-10.000
1.35000	-0.552	-101.442	-9.365	-9.365
1.40000	-0.603	-107.569	-8.989	-8.989
1.45000	-0.629	-113.529	-8.814	-8.814
1.50000	-0.630	-119.389	-8.809	-8.809
1.55000	-0.608	-125.202	-8.959	-8.959
1.60000	-0.566	-131.016	-9.258	-9.258
1.65000	-0.508	-136.865	-9.713	-9.713
1.70000	-0.439	-142.779	-10.338	-10.338
1.75000	-0.363	-148.777	-11.160	-11.160
1.80000	-0.286	-154.868	-12.218	-12.218
1.85000	-0.212	-161.055	-13.579	-13.579
1.90000	-0.146	-167.332	-15.350	-15.350
1.95000	-0.091	-173.684	-17.733	-17.733
2.00000	-0.050	179.909	-21.169	-21.169
2.05000	-0.025	173.470	-27.024	-27.024
2.10000	-0.017	167.026	-52.902	-52.902
2.15000	-0.023	160.600	-28.305	-28.305
2.20000	-0.042	154.213	-22.317	-22.317
2.25000	-0.071	147.879	-19.068	-19.068
2.30000	-0.105	141.606	-16.960	-16.960
2.35000	-0.141	135.390	-15.510	-15.510
2.40000	-0.174	129.223	-14.511	-14.511
2.45000	-0.201	123.086	-13.861	-13.861
2.50000	-0.217	116.951	-13.510	-13.510
2.55000	-0.221	110.783	-13.442	-13.442
2.60000	-0.211	104.542	-13.674	-13.674
2.65000	-0.187	98.178	-14.254	-14.254
2.70000	-0.152	91.641	-15.292	-15.292
2.75000	-0.110	84.873	-17.016	-17.016
2.80000	-0.068	77.822	-19.983	-19.983
2.85000	-0.036	70.437	-26.188	-26.188
2.90000	-0.027	62.686	-37.735	-37.735
2.95000	-0.060	54.558	-21.430	-21.430
3.00000	-0.151	46.079	-15.610	-15.610

Table J.5C

$\theta = 60°$

$Z_1 = Z_0 = 22.5\Omega$				
FREQ-GHZ	DB[S21] PS	ANG[S21] PS	DB[S11] PS	DB[S22] PS
1.00000	-0.023	-47.238	-50.576	-50.576
1.05000	-0.078	-57.307	-18.842	-18.842
1.10000	-0.215	-66.405	-13.562	-13.562
1.15000	-0.387	-74.611	-10.898	-10.898
1.20000	-0.561	-82.062	-9.301	-9.301
1.25000	-0.716	-88.908	-8.284	-8.284
1.30000	-0.840	-95.294	-7.630	-7.630
1.35000	-0.928	-101.342	-7.229	-7.229
1.40000	-0.979	-107.161	-7.020	-7.020
1.45000	-0.992	-112.839	-6.966	-6.966
1.50000	-0.971	-118.453	-7.052	-7.052
1.55000	-0.919	-124.066	-7.272	-7.272
1.60000	-0.840	-129.730	-7.629	-7.629
1.65000	-0.741	-135.490	-8.137	-8.137
1.70000	-0.628	-141.377	-8.819	-8.819
1.75000	-0.508	-147.414	-9.711	-9.711
1.80000	-0.388	-153.612	-10.868	-10.868
1.85000	-0.275	-159.970	-12.382	-12.382
1.90000	-0.177	-166.473	-14.407	-14.407
1.95000	-0.099	-173.095	-17.257	-17.257
2.00000	-0.045	-179.798	-21.742	-21.742
2.05000	-0.019	173.460	-31.756	-31.756
2.10000	-0.019	166.726	-30.633	-30.633
2.15000	-0.045	160.043	-21.675	-21.675
2.20000	-0.092	153.448	-17.583	-17.583
2.25000	-0.154	146.968	-15.030	-15.030
2.30000	-0.225	140.615	-13.270	-13.270
2.35000	-0.298	134.390	-12.010	-12.010
2.40000	-0.367	128.282	-11.105	-11.105
2.45000	-0.424	122.267	-10.478	-10.478
2.50000	-0.466	116.310	-10.084	-10.084
2.55000	-0.487	110.368	-9.902	-9.902
2.60000	-0.484	104.389	-9.929	-9.929
2.65000	-0.458	98.312	-10.183	-10.183
2.70000	-0.407	92.072	-10.704	-10.704
2.75000	-0.335	85.594	-11.573	-11.573
2.80000	-0.249	78.800	-12.945	-12.945
2.85000	-0.159	71.615	-15.143	-15.143
2.90000	-0.080	63.970	-19.033	-19.033
2.95000	-0.032	55.823	-28.998	-28.998
3.00000	-0.041	47.170	-25.414	-25.414

Table J.6A

$Z_1 = 32\Omega$, $\theta = 60°$

	$Z_0 = 45\Omega$			
FREQ-GHZ	DB[S21]	ANG[S21]	DB[S11]	DB[S22]
	PS	PS	PS	PS
1.00000	−0.042	−59.644	−23.151	−23.151
1.05000	−0.020	−67.227	−42.202	−42.202
1.10000	−0.028	−74.452	−26.843	−26.843
1.15000	−0.054	−81.350	−20.881	−20.881
1.20000	−0.089	−87.956	−17.899	−17.899
1.25000	−0.127	−94.314	−16.076	−16.076
1.30000	−0.161	−100.463	−14.883	−14.883
1.35000	−0.191	−106.444	−14.095	−14.095
1.40000	−0.212	−112.292	−13.595	−13.595
1.45000	−0.225	−118.040	−13.319	−13.319
1.50000	−0.230	−123.715	−13.229	−13.229
1.55000	−0.227	−129.342	−13.300	−13.300
1.60000	−0.216	−134.942	−13.523	−13.523
1.65000	−0.200	−140.531	−13.894	−13.894
1.70000	−0.179	−146.122	−14.417	−14.417
1.75000	−0.155	−151.726	−15.103	−15.103
1.80000	−0.130	−157.347	−15.971	−15.971
1.85000	−0.106	−162.991	−17.052	−17.052
1.90000	−0.083	−168.657	−18.394	−18.394
1.95000	−0.062	−174.344	−20.076	−20.076
2.00000	−0.046	179.950	−22.233	−22.233
2.05000	−0.033	174.229	−25.136	−25.136
2.10000	−0.025	168.499	−29.452	−29.452
2.15000	−0.021	162.763	−37.877	−37.877
2.20000	−0.020	157.025	−43.197	−43.197
2.25000	−0.023	151.286	−32.061	−32.061
2.30000	−0.028	145.547	−27.772	−27.772
2.35000	−0.033	139.806	−25.304	−25.304
2.40000	−0.039	134.057	−23.762	−23.762
2.45000	−0.044	128.294	−22.835	−22.835
2.50000	−0.047	122.504	−22.400	−22.400
2.55000	−0.047	116.676	−22.425	−22.425
2.60000	−0.045	110.794	−22.956	−22.956
2.65000	−0.040	104.838	−24.141	−24.141
2.70000	−0.034	98.790	−26.358	−26.358
2.75000	−0.028	92.629	−30.777	−30.777
2.80000	−0.026	86.333	−46.997	−46.997
2.85000	−0.030	79.885	−31.625	−31.625
2.90000	−0.045	73.266	−24.098	−24.098
2.95000	−0.076	66.467	−19.630	−19.630
3.00000	−0.131	59.484	−16.370	−16.370

Table J.6A

$Z_1 = 32\Omega$, $\theta = 60°$

$Z_0 = 53\Omega$				
FREQ-GHZ	DB[S21] PS	ANG[S21] PS	DB[S11] PS	DB[S22] PS
1.00000	-0.020	-64.923	-37.811	-37.811
1.05000	-0.037	-71.956	-23.837	-23.837
1.10000	-0.069	-78.667	-19.369	-19.369
1.15000	-0.108	-85.095	-16.885	-16.885
1.20000	-0.148	-91.281	-15.305	-15.305
1.25000	-0.184	-97.267	-14.251	-14.251
1.30000	-0.214	-103.090	-13.549	-13.549
1.35000	-0.236	-108.785	-13.104	-13.104
1.40000	-0.249	-114.384	-12.864	-12.864
1.45000	-0.253	-119.914	-12.797	-12.796
1.50000	-0.248	-125.399	-12.883	-12.883
1.55000	-0.236	-130.859	-13.114	-13.114
1.60000	-0.218	-136.310	-13.491	-13.491
1.65000	-0.195	-141.767	-14.019	-14.019
1.70000	-0.168	-147.237	-14.712	-14.712
1.75000	-0.141	-152.729	-15.593	-15.593
1.80000	-0.113	-158.245	-16.702	-16.702
1.85000	-0.087	-163.787	-18.097	-18.097
1.90000	-0.064	-169.352	-19.879	-19.879
1.95000	-0.046	-174.937	-22.235	-22.235
2.00000	-0.032	179.463	-25.571	-25.571
2.05000	-0.023	173.854	-31.096	-31.096
2.10000	-0.020	168.244	-49.322	-49.322
2.15000	-0.022	162.637	-33.858	-33.858
2.20000	-0.028	157.040	-27.376	-27.376
2.25000	-0.038	151.456	-23.922	-23.922
2.30000	-0.050	145.887	-21.665	-21.665
2.35000	-0.063	140.333	-20.078	-20.078
2.40000	-0.077	134.789	-18.941	-18.941
2.45000	-0.088	129.252	-18.142	-18.142
2.50000	-0.097	123.712	-17.624	-17.624
2.55000	-0.102	118.159	-17.359	-17.359
2.60000	-0.103	112.579	-17.343	-17.343
2.65000	-0.099	106.956	-17.590	-17.590
2.70000	-0.091	101.273	-18.144	-18.144
2.75000	-0.078	95.510	-19.088	-19.088
2.80000	-0.063	89.648	-20.588	-20.588
2.85000	-0.047	83.667	-23.008	-23.008
2.90000	-0.034	77.546	-27.394	-27.394
2.95000	-0.028	71.270	-40.640	-40.640
3.00000	-0.032	64.825	-30.459	-30.459

Table J.6B

$Z_1 = 32\Omega$, $\theta = 60°$

$Z_0 = 64\Omega$			
FREQ-GHZ	DB[S21]	ANG[S21]	DB[S11]	DB[S22]
	PS	PS	PS	PS
1.00000	-0.049	-69.404	-21.556	-21.556
1.05000	-0.087	-75.939	-18.024	-18.024
1.10000	-0.131	-82.199	-15.914	-15.914
1.15000	-0.174	-88.224	-14.527	-14.527
1.20000	-0.212	-94.054	-13.585	-13.585
1.25000	-0.244	-99.726	-12.949	-12.949
1.30000	-0.267	-105.275	-12.545	-12.545
1.35000	-0.280	-110.730	-12.325	-12.325
1.40000	-0.284	-116.119	-12.264	-12.264
1.45000	-0.279	-121.465	-12.345	-12.345
1.50000	-0.266	-126.788	-12.563	-12.563
1.55000	-0.247	-132.105	-12.918	-12.918
1.60000	-0.222	-137.428	-13.415	-13.415
1.65000	-0.193	-142.769	-14.070	-14.070
1.70000	-0.162	-148.134	-14.905	-14.905
1.75000	-0.131	-153.527	-15.956	-15.956
1.80000	-0.101	-158.950	-17.282	-17.282
1.85000	-0.075	-164.401	-18.977	-18.977
1.90000	-0.053	-169.877	-21.216	-21.216
1.95000	-0.036	-175.372	-24.376	-24.376
2.00000	-0.025	179.122	-29.524	-29.524
2.05000	-0.020	173.611	-44.006	-44.006
2.10000	-0.022	168.105	-33.778	-33.778
2.15000	-0.029	162.610	-26.660	-26.660
2.20000	-0.042	157.134	-22.943	-22.943
2.25000	-0.059	151.681	-20.491	-20.491
2.30000	-0.079	146.255	-18.723	-18.723
2.35000	-0.100	140.857	-17.398	-17.398
2.40000	-0.122	135.485	-16.391	-16.391
2.45000	-0.141	130.136	-15.634	-15.634
2.50000	-0.158	124.802	-15.084	-15.084
2.55000	-0.171	119.475	-14.719	-14.719
2.60000	-0.178	114.145	-14.526	-14.526
2.65000	-0.179	108.796	-14.503	-14.503
2.70000	-0.174	103.415	-14.658	-14.658
2.75000	-0.163	97.984	-15.012	-15.012
2.80000	-0.146	92.485	-15.601	-15.601
2.85000	-0.124	86.900	-16.489	-16.489
2.90000	-0.099	81.210	-17.788	-17.788
2.95000	-0.073	75.396	-19.719	-19.719
3.00000	-0.050	69.441	-22.790	-22.790

Table J.6B

$Z_1 = 32\Omega,\ \theta = 60°$

$Z_0 = 70\Omega$			
FREQ-GHZ	DB[S21] PS	ANG[S21] PS	DB[S11] PS	DB[S22] PS
1.00000	-0.077	-70.858	-18.722	-18.722
1.05000	-0.123	-77.164	-16.234	-16.234
1.10000	-0.170	-83.215	-14.623	-14.623
1.15000	-0.215	-89.052	-13.524	-13.524
1.20000	-0.254	-94.712	-12.765	-12.765
1.25000	-0.285	-100.230	-12.253	-12.253
1.30000	-0.306	-105.639	-11.934	-11.934
1.35000	-0.317	-110.966	-11.774	-11.774
1.40000	-0.319	-116.238	-11.753	-11.753
1.45000	-0.311	-121.475	-11.861	-11.861
1.50000	-0.296	-126.696	-12.092	-12.092
1.55000	-0.273	-131.918	-12.448	-12.448
1.60000	-0.246	-137.151	-12.937	-12.937
1.65000	-0.215	-142.407	-13.571	-13.571
1.70000	-0.181	-147.691	-14.370	-14.370
1.75000	-0.148	-153.008	-15.367	-15.367
1.80000	-0.115	-158.358	-16.612	-16.612
1.85000	-0.086	-163.741	-18.184	-18.184
1.90000	-0.061	-169.151	-20.223	-20.223
1.95000	-0.042	-174.585	-23.010	-23.010
2.00000	-0.028	179.967	-27.248	-27.248
2.05000	-0.021	174.510	-35.932	-35.932
2.10000	-0.021	169.054	-39.020	-39.020
2.15000	-0.026	163.606	-28.487	-28.487
2.20000	-0.037	158.173	-24.015	-24.015
2.25000	-0.053	152.762	-21.226	-21.226
2.30000	-0.072	147.375	-19.264	-19.264
2.35000	-0.093	142.014	-17.809	-17.809
2.40000	-0.115	136.679	-16.708	-16.708
2.45000	-0.135	131.366	-15.876	-15.876
2.50000	-0.153	126.069	-15.264	-15.264
2.55000	-0.167	120.779	-14.843	-14.843
2.60000	-0.176	115.487	-14.598	-14.598
2.65000	-0.179	110.179	-14.524	-14.524
2.70000	-0.176	104.840	-14.626	-14.626
2.75000	-0.166	99.453	-14.921	-14.921
2.80000	-0.150	94.000	-15.440	-15.440
2.85000	-0.130	88.462	-16.241	-16.241
2.90000	-0.105	82.819	-17.420	-17.420
2.95000	-0.080	77.051	-19.165	-19.165
3.00000	-0.056	71.141	-21.886	-21.886

Table J.6C

$Z_1 = 32\Omega, \; \theta = 60°$

		$Z_0 = 76\Omega$		
FREQ-GHZ	DB[S21] PS	ANG[S21] PS	DB[S11] PS	DB[S22] PS
1.00000	-0.101	-72.155	-17.268	-17.268
1.05000	-0.149	-78.301	-15.269	-15.269
1.10000	-0.197	-84.210	-13.928	-13.928
1.15000	-0.241	-89.920	-13.000	-13.000
1.20000	-0.278	-95.468	-12.358	-12.358
1.25000	-0.306	-100.889	-11.931	-11.931
1.30000	-0.324	-106.211	-11.677	-11.677
1.35000	-0.332	-111.463	-11.569	-11.569
1.40000	-0.331	-116.667	-11.592	-11.592
1.45000	-0.320	-121.844	-11.737	-11.737
1.50000	-0.302	-127.012	-12.002	-12.002
1.55000	-0.277	-132.185	-12.391	-12.391
1.60000	-0.247	-137.374	-12.911	-12.911
1.65000	-0.214	-142.589	-13.580	-13.580
1.70000	-0.179	-147.835	-14.419	-14.419
1.75000	-0.145	-153.116	-15.464	-15.464
1.80000	-0.112	-158.432	-16.770	-16.770
1.85000	-0.083	-163.780	-18.429	-18.429
1.90000	-0.058	-169.157	-20.604	-20.604
1.95000	-0.039	-174.557	-23.638	-23.638
2.00000	-0.026	-179.971	-28.464	-28.464
2.05000	-0.020	174.609	-40.390	-40.390
2.10000	-0.021	169.190	-34.723	-34.723
2.15000	-0.029	163.782	-26.827	-26.827
2.20000	-0.043	158.393	-22.891	-22.891
2.25000	-0.061	153.027	-20.323	-20.323
2.30000	-0.083	147.690	-18.474	-18.474
2.35000	-0.107	142.384	-17.081	-17.081
2.40000	-0.131	137.108	-16.010	-16.010
2.45000	-0.155	131.859	-15.188	-15.188
2.50000	-0.176	126.633	-14.569	-14.569
2.55000	-0.193	121.422	-14.125	-14.125
2.60000	-0.205	116.216	-13.839	-13.839
2.65000	-0.212	111.003	-13.706	-13.706
2.70000	-0.211	105.769	-13.726	-13.726
2.75000	-0.204	100.498	-13.907	-13.907
2.80000	-0.190	95.174	-14.270	-14.270
2.85000	-0.170	89.778	-14.849	-14.849
2.90000	-0.144	84.292	-15.699	-15.699
2.95000	-0.116	78.696	-16.919	-16.919
3.00000	-0.086	72.971	-18.693	-18.693

Appendix K

Derivation of the Input Signal in Figure 2.21

At the output of the gain block $-g$,

$$v = R_2 C_2 \frac{dW}{dt} + W$$

So, at the input of the gain block $-g$,

$$v' = -\frac{v}{g}$$

Therefore, at the input of the circuit in Figure 2.21,

$$
\begin{aligned}
u &= R_1 C_1 \frac{dv'}{dt} + v' \\[2mm]
&= R_1 C_1 \frac{d}{dt} \left[\frac{R_2 C_2}{g} \frac{dW}{dt} + \frac{W}{g} \right] - \frac{R_2 C_2}{g} \frac{dW}{dt} + \frac{W}{g} \\[2mm]
&= -\frac{T_1 T_2}{g} \frac{d^2 W}{dt^2} - \frac{T_1}{g} \frac{dW}{dt} - \frac{T_2}{g} \frac{dW}{dt} - \frac{W}{g} \\[2mm]
&= -\frac{T_1 T_2}{g} \frac{d^2 W}{dt^2} - \frac{T_1 + T_2}{g} \frac{dW}{dt} - \frac{W}{g}
\end{aligned}
$$

where

$$T_1 = R_1 C_1$$
$$T_2 = R_2 C_2$$

Therefore, looking right toward the circuit,

$$u = y_{ii1}(t)$$

and looking left toward the multiplier,

$$u = y_{om1}(t)$$

About the Author

M. A. Halim was educated in Bangladesh, England, and Holland. In 1967 he moved from Europe to Canada and has worked with various organizations, including Leigh Instruments Limited, Spar Aerospace Limited, Canadian Marconi Company, and Nortel. Most recently he was with SR Telecom, one of the leading manufacturers and global suppliers of wireless telephones. His activities there included different types of antenna design and solving various EMI/EMC problems, including the effect of EM radiation on humans, especially that from cell phones on human brains, which was detailed in his article in the December 1999 issue of *Microwaves & RF*. He has published many papers in various technical journals on different subjects and is the author of two patents. He developed the Halim phase shifter, a class of wideband phase shifters providing large phase shifts. He recently developed the Halim hybrids, a new class of 3-dB 180° hybrids, which is in the process of being patented by SR Telecom.

Index

automatic adjustment of weighter,
23–26
degree of, 46
manual adjustment of weighter, 23
matrix expressions, 204–6
principle, 22
two-source, two-output system, 46–50
Characteristic equation, 207
Characteristic roots, 256
Characteristic vector, 256
Closed-loop transfer function, 137–39
loop differential equation, 137–38
loop transfer function, 138–39
nonoscillatory two-pole feedback loop,
152
See also Feedback loop
Column matrix, 196, 255
Convergence time
defined, 139
feedback loop and, 139–53
as function of loop gain, 140–41
wide difference in, 141
Coupling coefficient, 174
bandwidth as function of, 175
variation of return loss and normalized
coupler spacing, 176
Covariance matrix, 203

Diagonal matrix, 256

Eigenvalues, 256
interference signal sources and, 225
physical significance of, 215–27
relation between signal powers and,
221–27
signal matrix in terms of, 216–17
Endfire array
defined, 4
illustrated, 6
See also Linear arrays

Feedback loop, 136–53
bode plot, 140
closed-loop transfer function, 137–39
convergence time and, 139–53
disabled, 59
high-frequency poles, 147–49
modal expressions, 214–15

modified, 141–47
multisource, multiport system, 214–15
nonoscillatory two-pole, 149–53
with phase detector, 126, 128
root locus, 140
stability, 136–37
total gain function, 49
Final output power, 20–22
leakage into AUX ports, 113–14,
249–50
multisource, two-output system,
95–97, 243–48
offset voltage effect, 123–24, 251–52
space noise effect, 129–30, 131–33,
253–54
three-source, three-output system,
87–88, 239–42
three-source, two-output system,
78–79, 235–37
two-source, multioutput system, 71–72
two-source, three-output system, 59,
231–34
two-source, two-output system, 45–46,
229–30
Final output signals, 20
multisource, two-output system, 93
offset voltage effect, 120–21
space noise effect, 129–30
three-source, three-output system,
84–85
three-source, two-output system, 76
two-source, multioutput system, 67
two-source, three-output system, 53,
55–56
two-source, two-output system, 39–41,
41–43

Gain, 216
loop, convergence time as function of,
140–41
at output port, 18
between signal source and AUX output
ports, 218
total function, 49
Gain matrix
characteristics of, 217–21
expansion of, 217–19

Recent Titles in the Artech House Antennas and Propagation Library

Thomas Milligan, Series Editor

For further information on these and other Artech House titles, including previously considered out-of-print books now available through our In-Print-Forever® (IPF®) program, contact:

Artech House
685 Canton Street
Norwood, MA 02062
Phone: 781-769-9750
Fax: 781-769-6334
e-mail: artech@artechhouse.com

Artech House
46 Gillingham Street
London SW1V 1AH UK
Phone: +44 (0)20 7596-8750
Fax: +44 (0)20 7630 0166
e-mail: artech-uk@artechhouse.com

Find us on the World Wide Web at:
www.artechhouse.com